FOREWORD

The Committee for Fisheries considered various aspects at the interface between the coastal zone and the fishing and aquaculture interests in its programme of work for 1995 and 1996. At its 77th Session in March 1996, the Committee suggested that the conclusions of this work together with the country case studies submitted to the activity be made available to the public. This work is published on the responsibility of the OECD Secretary-General.

90 0345681 1

RECONCILING PRESSURES
ON THE COASTAL ZONE

Fisheries and Aquaculture

ORGANISATION FOR ECONOMIC CO-OPERATION AND DEVELOPMENT

ORGANISATION FOR ECONOMIC CO-OPERATION AND DEVELOPMENT

Pursuant to Article 1 of the Convention signed in Paris on 14th December 1960, and which came into force on 30th September 1961, the Organisation for Economic Co-operation and Development (OECD) shall promote policies designed:

- to achieve the highest sustainable economic growth and employment and a rising standard of living in Member countries, while maintaining financial stability, and thus to contribute to the development of the world economy;
- to contribute to sound economic expansion in Member as well as non-member countries in the process of economic development; and
- to contribute to the expansion of world trade on a multilateral, non-discriminatory basis in accordance with international obligations.

The original Member countries of the OECD are Austria, Belgium, Canada, Denmark, France, Germany, Greece, Iceland, Ireland, Italy, Luxembourg, the Netherlands, Norway, Portugal, Spain, Sweden, Switzerland, Turkey, the United Kingdom and the United States. The following countries became Members subsequently through accession at the dates indicated hereafter: Japan (28th April 1964), Finland (28th January 1969), Australia (7th June 1971), New Zealand (29th May 1973), Mexico (18th May 1994), the Czech Republic (21st December 1995) and Hungary (7th May 1996). The Commission of the European Communities takes part in the work of the OECD (Article 13 of the OECD Convention).

Publié en français sous le titre :

RÉCONCILIER LES PRESSIONS SUR LES ZONES CÔTIÈRES
Pêcheries et aquaculture

TABLE OF CONTENTS

PART I: RESOURCE USER CONFLICTS IN THE COASTAL ZONE 7

Summary and Observations 9

The aboriginal fisheries strategy: A Canadian approach to
coastal resource use conflicts 15

Strategies to resolve fishery management problems in the south-east arising
from competition between commercial and recreational fishermen 21

Conflict between shipping and fishing in the Finisterre area 31

Short history of coastal zone planning in Norway, with
emphasis on the Municipality of Vega 35

The effects of dumping waste from titanium dioxide production
into the Helgoland Bight 43

Impact of development and waste disposal in the coastal
areas on fishing grounds 49

PART II:COASTAL ZONE RESPONSES TO CHANGING FISHING
POSSIBILITIES 59

Summary and Observations 61

The Atlantic groundfish strategy 73

Revilatisation of fishing communities in Japan............................. 85

A description of the Norwegian fisheries policy objectives and
the effects the measures have had 99

PART I

RESOURCE USER CONFLICTS IN THE COASTAL ZONE

SUMMARY AND OBSERVATIONS

Introduction

When the Committee for Fisheries decided to embark upon a study of the coastal zone, two main parts were identified: coastal resource user conflicts and coastal zone responses to changing fishing possibilities.

The following summarises the main points which have been drawn from Part I concerning resource user conflicts in the coastal zone. The observations are based on the reports which some Member countries submitted and on the discussions during Committee meetings.

The evidence

Background

In pursuing its activity on the coastal zone, the Committee has discussed various aspects of user conflicts based on six country case studies[1] provided by Member countries. These case studies consider a wide range of problems of resource user conflicts which may be grouped in three major categories:

---

[1] Canada: "The aboriginal fisheries strategy: A Canadian approach to coastal resource use conflicts"; United States: Strategies to resolve fishery management problems in the South-East arising from competition between commercial and recreational fishermen"; Norway: "Short history of coastal zone planning in Norway, with emphasis on the Municipality of Vega"; Spain: "Conflict between shipping and fishing in the Finisterre area"; Germany: "The effects of dumping waste from titanium dioxide production into the Helgoland Bight"; Japan: "Impact of development and waste disposal in the coastal areas on fishing grounds".

- conflicts of interest stemming from allocations to various user groups from a common property resource/fisheries (Canada, United States);

- conflicts between user groups stemming from the use of the same space/fishing grounds/harbour installation/sea lanes (Norway, Spain);

- conflicts between user groups due to one user group polluting the sea (Germany, Japan).

The point of departure for most of the discussion in the Committee and in the case studies has been the recognition that user conflicts may lead to coastal zone degradation, which is costly to society. In the case of fisheries, conflicts may be detrimental to fisheries habitats or could have a direct negative impact on the fish stocks.

User groups tend to use the resource without taking into account the possible externalities their practices have on other interested parties in the coastal zone. A parallel is found in the common property nature of fisheries. The total use of the coastal zone may exceed its carrying capacity when either property rights are not assigned, or other appropriate regulatory mechanisms, which internalise the externalities, are not introduced.

A range of conflicts are present in the coastal zone and in the area which it encompasses. Commonplace examples include the conflicting use of space (e.g. interference with commercial vessels lanes by pleasure boats or fishing vessels; in this regard see the Spanish case study on a traffic separation scheme in the Finisterre Area), pollution (vessel paint TBT's, anti fouling and its influence on shellfish) and allocations (commercial fishermen vs. sports fishermen). Less obvious cases are mineral exploitation, e.g. oil exploration, and sand dredging. However, common to all of them is that the conflicts appear because of the lack of property rights in some form or lack of management arrangements for the use of the coastal zone and the allocation of the resources.

The coastal zone and the sea have for long been publicly owned areas with free access and use. The introduction of some form of licensing and/or fees, reflecting the value of the use of the coastal zone and the sea, could be envisaged. It does not appear from the exercise that Member countries have instituted any form of tradeable/market based user rights for the coastal zone. Rather, and as shown in the country case studies, regulatory interventions are mostly used.

The various case studies evidence that under certain circumstances it is possible to contain the conflicts, or, through the creation of the appropriate institutions and regulations, diffuse potential conflicts. This has especially been the case in the studies which have dealt with the allocation of fishing or user rights. Success in the cases which have not introduced user rights has been less evident. Some plausible reasons for success or lack of success are developed in the following.

Recognition of the problems

Recognition of the problems in the coastal zone by all interested groups is a major step towards the resolution of potential conflicts. However, more often than not, the user groups dependent on the coastal zone work as dispersed units making it difficult for them to collaborate and co-ordinate activities. Hence it is important to have a public authority to analyse the coastal zone in order to develop a comprehensive policy for its development. See in this regard a previous study "Coastal Zone Management: Integrated Policies" (OECD, 1993).

The Canadian case study provides evidence of the benefits of consulting all parties to the conflicts in the coastal zone to identify a common ground for conflict resolution. In Norway local authorities are heavily involved through consultation procedures with the fisheries management and environmental authorities in the planning and use of the coastal zone for aquaculture development. The case study by Norway reports on the development of coastal zone planning with particular emphasis on the accommodation of increased aquaculture activities.

In a report by the Australian Parliament entitled "The Injured Coastline" (April 1991) the point is made that, in spite of having recognised the problems of the coastal zone during various hearings, conferences, inquiries etc. no comprehensive public action is taken. With a view to remedy this situation, a comprehensive coastal zone inquiry on the management and use of coastal zone resources was undertaken and finalised in 1993; a final report of this work was published by the Australian Resource Assessment Commission in November 1993. The Report concludes, inter alia, that a national approach to coastal zone management is essential because no single entity of Government can manage the coastal zone alone, that the coastal zone is a significant national issue of public concern, that the socio-economic development of the coast is of utmost importance and that international obligations necessitates co-ordination between various parts of the Government.

Generation of information

Generation of information by the regulatory agencies on the various conditions in the coastal zone is important for assessing the scale of the problems encountered. Various areas need to be analysed and quantitative and/or qualitative data should be collected on the physical environment, eco-systems, social characteristics of the various user groups, the economic importance of the user groups, etc. Such procedures, however, are costly when followed on a comprehensive basis, partly because the institutions are not present to collect this type of information. Hence, in general, data collection occurs on an ad hoc basis in response to a particular problem.

Also, due to the costs involved, data and information collection tends to be ex-post to the problems encountered, i.e. conflicts have occurred and relevant information is collected and assessed after the fact. This is typical of most coastal resource user conflicts as identified in the Recommendation of the OECD Council on Integrated Coastal Zone Management (reproduced in "Coastal Zone Management: Integrated Policies", OECD, 1993). In this regard little progress seems to have been achieved since the adoption of the recommendation. There continues to be a need for the integration and harmonisation of sectoral policies affecting the coastal zone management and resource usage. The development of indicators for the monitoring of coastal zone activities and processes should be pursued as has been the case with rural indicators.

The discussion in the Committee for Fisheries evidenced that information and data often are lacking and that authorities may take decisions based on limited data. Some Member countries (e.g. Norway and Australia) have adopted a more comprehensive approach which involves all levels of decision making and data collection. The Commission of the EU undertook a comprehensive Community wide information and data gathering on the coastal zone in 1992; the objective of this undertaking was to quantify the socio-economic importance of the fishing sector in the coastal zone including an evaluation of the coastal communities' dependence on fisheries.

Continuous monitoring and evaluation

A continuous monitoring and evaluation of the coastal zone and the different economic sectors, which co-exist there, is fundamental. As outlined in the "Strategies to resolve fisheries management problems in the south-east arising from competition between commercial and recreational fishermen" by

R.L. Schmied, NMFS, United States, commercial fisheries in the south-east United States have changed rapidly during the 70s; new types of fishing gear, etc. have the potential for adding new dimensions to conflicts. Furthermore the population in the south-eastern region of the United States grew 45 per cent between 1970 and 1988, adding pressure to the coastal area. Clearly, the monitoring of such developments is of prime importance and underlines the need for establishing relevant data and information bases.

As has been proven for other cases, where the management of a common property resource is being considered, the observance of the rules of the game is best secured by early public participation in the planning and choice of instruments. The Canadian case study has been particularly interesting in this regard. One component in the Canadian "Aboriginal Fisheries Strategy" calls for local participation as the issue of licences is communal and not individual.

The choice of regulatory mechanism also impinges on the user groups' willingness to observe the rules of the game. The Canadian case study clearly shows that when interest groups are approached and consulted, and a common plan developed and agreed upon, conflict resolution, in this particular case due to allocations of fishing rights, can be achieved.

Pollution problems may have serious effects on the fishing industry, including the contamination of fish and fish products. The Japanese case study evidences that discharges into the sea are causing damage to the fishing industry. This is a problem which also could have serious marketing effects on the fishing industry throughout Member countries, as demonstrated, for example, by the German nematode experience. In a recent report published by UNESCO in the IOC Ocean Forum series (Coastal Zone Space: Prelude to Conflict? by Edward D. Goldberg, UNESCO, 1994), reference is made to the rapidly developing tourism industry and the discharges of pathogens into the sea, with possible negative effects on fish and shellfish. The German case study provides evidence on the linkage between dumping harmful substances (e.g. heavy metals) and fish diseases.

Assessment of the costs of user conflicts

Assessment of the costs of user conflicts in Member countries is not done on a comprehensive basis. Too often, assessments of the costs involved in not pursuing integrated coastal zone management programmes which could reduce or eliminate the resource user conflicts are not carried out. Such lack of action may also increase the burden at a later stage.

Other

There are other less apparent areas which need to be addressed in the coastal zone, including proper institutional mechanisms, the type of regulatory framework for conflict resolution i.e. administrative regulation vs. market mechanisms, and the level of decision making.

THE ABORIGINAL FISHERIES STRATEGY: A CANADIAN APPROACH TO COASTAL RESOURCE USE CONFLICTS[1]

Overview

At first blush, the resource use conflict in the Pacific salmon industry on Canada's west coast appears to be simply the classic paradigm of multiple interests seeking access to a limited resource. If that were the case, the Pacific salmon industry might easily lend itself to an economics-based model of conflict resolution, using tools such as tradable user allocations. The factor which prevents this neat application is the aboriginal interest in the resource and the unique legal rights (the nature of which are only partially determined) attached thereto. Further, the existence of the aboriginal right in this and other resources, and the conflict created by the positioning of all interests competing for a piece of the proverbial pie is just one component of a much larger socio-economic, legal issue, namely the quest of Canada's aboriginal peoples for self determination and recognition of their aboriginal and treaty rights. However, not surprisingly, interim issues, such as the conflict created by the competing goals of fisheries conservation and resource allocation where there is an aboriginal interest, arise and must be managed.

Canada's response to this interim fisheries management issue is the Aboriginal Fisheries Strategy (AFS), a program which utilises an economics-based conflict resolution model as one of its components. This paper will look at the AFS, the context from which it arose, and the results of the application of an economics-based conflict resolution model known as the Salmon Licence Retirement Program.

Background

The coastal waters and rivers of the province of British Columbia on Canada's west coast are home to the valuable Pacific salmon resource. Three

---

[1] This paper has been prepared by the Native Affairs Directorate, Fisheries and Oceans, Government of Canada.

sectors seek access to salmon, the largest of which is the commercial salmon industry. It supports approximately 13 500 fishermen and 9 400 processing plant employees, and is allocated approximately 94 per cent of the annual salmon allocation. There is also a thriving and influential recreational fishery which relies on British Columbia's spectacular rivers and scenery to attract almost 800 000 sport fishers who contribute $1.2 billion in value-added dollars to the Canadian economy annually. The annual salmon allocation to the recreational sector is about 3 per cent of the Total Allowable Catch (TAC). The third sector is composed of British Columbia's aboriginal peoples, approximately 155 200 in number. For centuries they have looked to salmon as a mainstay of their economies, cultures and nutritional requirements. The annual salmon allocation to the recreational sector is about 3 per cent of the TAC.

The unique status of aboriginal peoples with respect to access to the fisheries

Until 1990, the Department of Fisheries and Oceans acknowledged the dependence of aboriginals on salmon through the issuance of food fish licences, but these licences were issued in a manner which did not necessarily include meaningful consultation with aboriginal peoples. Conflicts arose over issues such as allocations, gear restrictions, and the Department of Fisheries and Oceans' prohibition on sale of aboriginal food fish. Increasingly, aboriginals engaged in protest fisheries and blockades to register their opposition to the Department of Fisheries and Oceans' restrictions. The Department of Fisheries and Oceans, by necessity, responded with numerous and expensive enforcement actions, that, due to the evolving jurisprudence surrounding aboriginal rights, frequently were unsuccessful and provided no real deterrent value. This conflict aggravated long-standing tensions between the aboriginal and non-aboriginal harvesting sectors which are manifested annually through intense competitive positioning over salmon allocations.

As stated, the conflict over Pacific salmon allocations is only one aspect of the much bigger issue of aboriginal self determination, which in turn involves questions of aboriginal rights, self government, and land claims. Within the last twenty years, there has been significant movement by the federal, provincial and aboriginal governments to resolve these questions, through the courts and through modern treaty-making processes. Much work remains to be done because along with the other issues, there are vast areas of Canada in which treaties have not been completed. In British Columbia in particular, most of the province is subject to some twenty-two land claims. The treaty process is only now getting under way and it will take some time before these claims are resolved. In addition to the treaty process, British Columbia aboriginal groups have been utilising the courts to advance their interests, with some notable successes.

The 1990 landmark decision of the Supreme Court of Canada (Canada's highest appellate court) in R. v. Sparrow is one such success. Sparrow recognised that when aboriginal peoples prove they have traditionally fished in an area, they have a *constitutional* right (Constitution Act, 1982, section 35) to fish for food, social and ceremonial purposes (the "aboriginal fishing right"). Further, where allocation is concerned, this aboriginal fishing right takes priority over all other users (such as the commercial and sports sectors) after provision for conservation. Sparrow held that the Department of Fisheries and Oceans had to accommodate the aboriginal fishing right in managing the fisheries and would have to justify any infringement of it. The justification test provided by the Court was one elucidated in earlier Supreme Court cases and included a requirement to consult with aboriginals affected by decisions such as those on allocation and enforcement.

Sparrow did not address the issue of whether the aboriginal fishing right include the right to sell the fish. For years, aboriginal people have been selling their "food" fish. Canada has contended that this is an illegal act because the fish was not caught under the authority of a commercial licence. The sale of aboriginal "food" fish has further fuelled the resentment of commercial fishermen because they are undercut by the lower price of the aboriginal catch. aboriginal leaders have forced the issue by taking a series of commercial sale cases to the British Columbia courts. These cases have reached the Supreme Court of Canada and will be heard and decided over the next two years.

The claims being advanced by aboriginal people in British Columbia have the potential to affect a number of stockholders. These included all stockholders in natural resource industries such as fishing, forestry and mining, land developers, and the tax payer who will have to bear the brunt of any financial settlement. As a result, although there is wide spread sympathy for the plight of aboriginal people who often live in poverty and hardship, there is also considerable apprehension over the price that will have to be paid to settle aboriginal peoples claims. When the uncertainty associated with the claims process is applied to the competitive fisheries context, the result is a highly charged and volatile environment for some of the non-aboriginal stockholders who fear they have everything to lose by an increased presence by aboriginal people in the fishery.

The aboriginal fisheries strategy (the "AFS")

Sparrow set the stage for a redefinition of Canada's relationship with aboriginal people regarding aboriginal fishing. The decision presented Canada with an immediate necessity to change its aboriginal fisheries policy so that it harmonised with the realities of a partially defined and evolving aboriginal right, pending final resolution of the issues through contemporary treaty-making processes which are ongoing or further court judgements. The result of this harmonisation is the AFS - Canada's policy response to Sparrow, currently about to begin the third year of a seven year mandate. The starting point for the AFS is the conservation and preservation of the resource and accordingly, the AFS emphasises and maintains the ultimate accountability of Canada for achieving this objective. It seeks to ensure the right to fish for food, social and ceremonial purposes through consultations with aboriginal groups which then lead to allocations in priority to all other users. The concept of co-operative management of the aboriginal fishery is introduced through consultations and through funding of fisheries-related economic development opportunities for Aboriginal communities.

The AFS is being implemented through the following program delivery components:

- annual negotiation with aboriginal groups of fisheries management agreements ("Agreements") and associated grants and contribution arrangements on a spectrum of activities which provide for co-operative management of the aboriginal fishery, habitat restoration projects, and fisheries-based economic development;

- the issuance of communal (versus individual) licences to aboriginal groups for implementation by local aboriginal fishing authorities, the terms of which reflect the above annual negotiations;

- a test program of commercial fishing licence purchase and retirement to facilitate the transfer of allocations to aboriginal fisheries to increase their commercial involvement in the fisheries, e.g., sale of aboriginal food fish, while ensuring minimum disruption to the industry; and,

- a co-operative participatory process involving all stockholders in fisheries.

Given the history of the user conflict over Pacific salmon, it was not surprising that the AFS has been controversial. Some of the commercial and sports sectors have been critical of the AFS, saying that the AFS would bring new pressures on the coastal waters and rivers in which the salmon are found and that the resource could not bear these pressures. Opponents to the AFS made dire predictions of the collapse of the Pacific salmon stocks as a result of the program. Their real concerns, broadly speaking, were first, that there was not enough salmon for all users, and second, that aboriginals should not be involved in co-management of the aboriginal fishery.

The policy, however, seeks to deal with these objections and thus minimise conflict and disruption in the fisheries sector through the program delivery components described above. In particular, an all-stakeholders process and a voluntary Salmon Licence Retirement Program are the vehicles used to meet these objectives. The all-stakeholders process is operated by industry participants from the commercial, sports and aboriginal sectors. The Department of Fisheries and Oceans' involvement in this process is minimal and is confined to funding it. The other program delivery component, the voluntary Salmon Licence Retirement Program, is discussed below.

Through the AFS, the Department of Fisheries and Oceans has been able to substantially increase the involvement of aboriginal peoples in British Columbia in fisheries co-management while at the same time ensuring record salmon returns in 1993. With respect to the conflict resolution aspect of the AFS, there is no doubt that it has lead to a better balancing of the various interests, but this balancing has been painful. Opposition to the AFS by some non-aboriginal sectors remains strong and the search for a method to ease these concerns continues.

Economic criteria as a basis for conflict resolution

The other AFS program component designed to meet the interests of commercial fishermen is the voluntary Salmon Licence Retirement Program. This is the portion of the AFS which uses economic criteria as a basis for conflict resolution. It is a two year (1992-1993), $7 million pilot licence retirement program for full fee commercial salmon licences. The purpose of the program was to test ways of reducing catching power in the commercial fleet when commercial fishing opportunities are transferred to aboriginal groups under the AFS. The Department of Fisheries and Oceans asked an industry board to develop proposals and analysis of options for compensation of licence holders who retire from the salmon industry. This was done and the program was carried out with licence bought and allocations redistributed accordingly. By an independent evaluation, there was no arbitrary loss to the commercial

19

sector through the program. Yet some opposition continues based on an expectation of potential loss stemming from future transfer through other processes such as the treaty process or court decisions. Continuing uncertainty fuels unabated fears by a segment of the commercial sector that massive displacement may still occur.

Conclusion

As stated in the *Overview* section of this paper, the difficulty with creating a conflict resolution mechanism for resource user conflict over Pacific salmon is the anxiety generated by the evolving definition of aboriginal rights to fish and of the lengthy processes that will be necessarily engaged to reach lasting solutions. This does not mean that an economic model does not have a role to play; Canada's experience with the Salmon Licence Retirement Program demonstrates that it represents an opportunity for government to intervene in the market-place to redistribute resources without causing harm to other stockholders. However, what our experience does suggest is that any issue having the complexity and socio-political impact attached to the resolution of land claims, self government issues, and aboriginal rights will not be resolved by an approach based solely on an economic model, but rather on a multi-disciplinary approach which is as cognisant of the human factors as it is of the theoretical factors.

STRATEGIES TO RESOLVE FISHERY MANAGEMENT PROBLEMS IN THE SOUTH-EAST ARISING FROM COMPETITION BETWEEN COMMERCIAL AND RECREATIONAL FISHERMEN[1]

Summary

The underlying problem resulting in conflicts between recreational and commercial fishermen in the Southeast Region is continued depletion of limited fishery resources through overfishing. The resources cannot sustain both groups' combined harvest demands.

In addition, a decline in water quality and aquatic habitat, associated with growing demands for seafood and sportfishing opportunities, are chiefly responsible for reduced fisheries production and user conflict. The conflicts are further exacerbated by changing social, economic, and political forces. These conditions are expected to continue in the future.

Current problems between and among both groups have arisen for the following reasons:

- direct harvest competition;

- commercial bycatch of recreationally important species;

- directed commercial fishing for forage species;

- localised overfishing (time/area conflicts);

- fishing and related gear impacts on aquatic habitat or protected species; and

- unethical fishing practices.

---

[1] This paper has been prepared by *Ronald L. Schmied,* National Marine Fisheries Service, Southeast Regional Office, St. Petersburg, Florida, United States.

Consequently, future management of marine fisheries in the Southeast will have to address the issues of: 1) overfishing, 2) protection and restoration of fisheries resources and associated habitat, and 3) allocation of fish resources among recreational and commercial fishermen in an increasingly competitive environment.

Background

Population growth and habitat loss

The 8 state coastal south-eastern region of the United States, including Puerto Rico and the US Virgin Islands contains about 55 per cent of the total coastline of the contiguous United States. Substantial fishing infrastructure exists along the region's shoreline providing the physical support for recreational and commercial fishing operations. This assemblage includes numerous coastal marinas, fishing piers, private and commercial recreational boats, and miles of fishable beaches, bridges, jetties, and artificial reefs.

The region's coastal wetlands, containing all of the nation's mangrove swamp wetlands and most of it's seagrasses, comprises about 83 per cent of the coastal wetlands in the coterminous United States. Since approximately 96 per cent of commercial and 70 per cent of recreational fishery resources in the Southeast are estuarine dependent, the protection and restoration of the region's estuarine systems are of paramount importance to it's fisheries production.

During 1988, about 2.3 billion pounds of fish and shellfish (worth about $881 million) were commercially harvested in the region. The combined regional seafood harvest represents 31 per cent of the volume, and about 25 per cent of the value of the fishery resources harvested in the United States. That year, nearly 7.5 million recreational fishermen took 40 million fishing trips, and caught about 194 million fish in the Southeast. Also during that year, the Southeast accounted for over 40 per cent of the nation's saltwater anglers, 56 per cent of all the trips, 50 per cent of the catch by numbers of fish, and over 55 per cent ($3.4 billion) of all direct expenditures made nationally by saltwater fishermen.

Despite their great importance, coastal wetlands nation-wide are being depleted at an annual rate of 20 000 acres. In the Southeast, it has been estimated such losses accounted for 84 per cent of the wetland losses nation-wide. One of the reasons for these dramatic wetland losses is the rapid population growth and associated coastal development that has occurred in the region. Between 1970 and 1988, the regional population grew 45 per cent, and by 2010, another 28 per cent population increase is projected. Much of this growth has occurred along the coastal margin placing additional development pressure on critical fisheries habitat. While researchers have not quantified the cause and effect relationship, it is clear that estuarine habitat loss and water quality impacts have contributed to reduced fisheries production and user competition and conflict. This fact should therefore be factored into conflict resolution strategies.

Commercial fisheries in transition

Commercial fisheries have undergone dramatic changes since 1970. Significant growth occurred in many sectors (shrimp, reef fish, coastal and oceanic pelagics) resulting, in part, from fisheries development initiatives sponsored by co-operative government/industry programs. This growth peaked around 1979 and began stalling out in the early 1980's in the face of upward spiraling inflation and interest rates. Profit margins for many commercial businesses declined steeply due to rapid increases in operating costs (especially fuel and insurance), over-capitalisation in some sectors (e.g. shrimping), reduced access to some species and fishing grounds, reduced landings due to fluctuations in natural production, reduced allocations due to overfishing and competition with recreational fishermen, and growing competition from imported seafood products.

In response to this rapidly changing environment, fishermen shifted to less utilised species and more efficient fishing gear (traps, bottom longlines, gillnets, purse seines and roller trawls). These changes led to increased competition and conflict among commercial fishermen (e.g. hook and line vs. gillnet fishermen), and between commercial and recreational fishermen. Major controversies erupted over allocation and management of numerous species like king and Spanish mackerel, red drum, billfish, and reef fish. Allocation decisions became less and less favourable to commercial interests as fishery resources became increasingly stressed and as recreational interests became better organised and more effective in state and federal fishery management programs.

Growth of marine recreational fisheries

In contrast to the commercial sector, marine recreational fishing grew through the 1970s and 1980s in spite of declines in the abundance of many target species and competition with the commercial sector. This is partially explained by strong regional population growth and tourism-based economies of many coastal communities. Future demand for marine recreational fishing, which is projected to increase by 45 per cent between 1985 and 2025, will also be driven by continued population expansion and influenced by general economic, environmental and fisheries conditions.

In addition to increases in the numbers of fishermen, technological advances in boat construction and propulsion, fishing tackle and techniques, electronics, and public education have all increased the "fishing power" of recreational enthusiasts. This "high tech" transition, paralleling commercial fisheries growth, has placed more pressure on fish stocks and increased competition between commercial and recreational fishermen. Consequently, between 1980 and 1990, there has been an explosion of angling regulations designed to reduce fishing pressure and user conflicts.

Also of note, during the 1980's, marine recreational fishermen used their new-found political clout to influence fisheries conservation and allocation decisions to the point where 10 finfish species have been designated as gamefish at state and/or federal levels.

Management issues

Commercial and recreational conflicts have been grouped into the following categories based on the principal cause of the conflict.

Direct harvest competition

Most regionally important fishery resources are targeted by both recreational and commercial interests. This competition has often evolved into bitter conflict as target species declined due to overharvest or natural fluctuations, with each faction blaming the other for problems in the fishery. Regulatory institutions are often faulted for failing to identify, prevent or resolve the problem(s). Conflict resolution is often delayed due to inadequate harvest or biological data, political infighting, procedural hang-ups, and disagreement on management measures.

Examples of this type of conflict abound and include existing and developing competition for reef fish (snapper, grouper, amberjack, sea bass, etc.), coastal pelagics (mackerel, cobia, dolphin, etc.), spiny lobster, red drum, and shark.

Commercial bycatch of recreational species

Presently, the enormous problem of bycatch of important finfish in the shrimp trawl fishery has become such a controversial issue that the US Congress has become directly involved and the National Marine Fisheries Service (NMFS) is mobilising a major co-operative initiative to resolve this problem.

The magnitude of this problem becomes apparent when one compares the (1972-1985) average annual shrimp trawl bycatch of 8 recreationally significant finfish species with the (1981-85) average annual recreational catch of those species. The shrimp bycatch was over 37 times greater in terms of number of fish caught.

Other bycatch problems include billfish and shark bycatch in the tuna and swordfish longline fisheries, mixed bycatch in drift gill net and fish trap fisheries, and hook/release mortality in recreational fisheries.

Gear and/or fishing modifications must be developed to reduce bycatch in the entire industry.

Commercial harvest of forage species

Recreational and commercial fishermen continually criticise NMFS' efforts to develop commercial fisheries for "under-utilised" species which serve as prey for higher valued predator species. These criticisms focus on our lack of understanding of predator-prey relationships and trophic level dynamics. It also reflects concern that without proper controls, commercial interests would decimate forage stocks, thereby harming predator species.

An example of such concern deals with overharvest of croaker stocks in the Gulf. Industrial bottomfish and food fish fishery, coupled with substantial croaker bycatch in the shrimp trawl fishery, resulted in an estimated 85 per cent decrease in croaker biomass between 1972 and 1987. Recreational and commercial hook and line fishermen were highly incensed over this over-exploitation of croaker.

Similar concern surfaced as efforts were launched in the Southeast to develop or expand fisheries for other forage species (Spanish sardine, Atlantic bumper, mullet, etc.). Past controversies regarding these fisheries have not been severe. However, recreational fishing leaders are giving more importance to these ventures because of growing concern for the depressed condition of many marine fish populations.

Localised overfishing

Fishing improvements, linked with the increasingly competitive nature of the region's fisheries and decline of target species populations, have precipitated numerous cases of localised overfishing. Localised overfishing occurs when commercial fishermen significantly reduce the availability of target species to other fishermen. While the local harvest may not constitute overfishing, it results in unequal resource shares and user confrontations.

The harvest of king and Spanish mackerel are examples of localised overfishing. Because of the migratory nature of these species, fishermen have access to the resource only during those times when the fish are moving through specific management areas. At times, total allowable catch has been harvested "upstream" and the fishery closed, or bag limits reduced to zero, before "downstream" fishermen have had an opportunity to harvest the resource. Remedy of the problem has taken the form of altering start and close of the fishing season, instituting equal or differential bag limits, and establishing recreational and commercial quotas.

Other local overfishing conflicts have emerged over deployment of fish traps on reefs, and use of nets, bottom longlines, purse seines, roller trawls, spear guns and "bang sticks" on target species.

Fishing Impacts on Habitat and Protected Species

Concern is increasing over damage to fisheries habitat as recreational and commercial groups become more sensitive in the protection of marine resources. Examples include direct harvest and gear damage to coral reefs; trawl impacts on seagrasses, other resources, and water quality; entanglement of sea birds and living marine resources in lost or discarded fishing gear; and propeller damage to sea grasses, marine mammals and turtles.

These controversies are side issues in the larger recreational and commercial fishing disputes.

Unethical fishing practices

Conflicts arising from unethical fishing practices are serious, but these are generally side issues in larger, more basic, conflicts. They serve to heighten existing tension and mistrust among fishermen and resource managers, and tend to confound resolution of larger, more important, conflicts. Examples include real or perceived under or over-reporting of catch, waste of catch, regulatory violations, mislabelling of seafood products, littering and lack of courtesy and respect for other fishermen or resource users.

Strategies for addressing each problem category

Direct harvest conflict

- Aggressively implement the Administration's "Zero Net Loss" habitat policy to protect, enhance and restore fisheries habitat to achieve the highest possible rate of fisheries production.

- Upgrade stock assessment capabilities to ensure accurate, timely and forward looking assessments for key species. Upgrade co-operative fishery dependent data-collection programs to obtain accurate and timely estimates of recreational and commercial fishing catch, effort, and related social/economic indices needed for fisheries allocation, monitoring and valuation on regional and sub-regional levels.

- Develop fishery independent data collection systems to support stock assessment and related fishery management activities.

- Develop/improve fishery management techniques to more effectively, efficiently and less intrusively control recreational and commercial fisheries effort and catch.

- Streamline management systems to ensure effective and timely responses to resource and user-based fishery management problems.

- Increase compliance with fishery management measures through improved regulatory, enforcement and education efforts.

- Advance quality and performance of scientific, management and enforcement staff through specialised training, education and performance incentives.

- Develop innovative funding mechanisms to support program enhancements.

Bycatch conflicts (indirect harvest competition)

- Develop and employ improved monitoring programs to characterise and quantify bycatch in commercial and recreational fisheries.

- Greatly expand the region's co-operative conservation engineering program to develop or improve fishing gear and fishing techniques that minimise or eliminate bycatch and/or bycatch mortality in commercial and recreational fisheries.

- Implement and enforce management measures to minimise or eliminate bycatch and/or bycatch mortality.

- Establish an expanded educational outreach program that encourages and convinces recreational and commercial fishermen to eliminate fishery resource waste.

Directed commercial harvest of forage species

- Implement an expanded research effort to characterise and quantify prey-predatory relationships and trophic-level dynamics for key species.

- Improve and expand stock assessment capabilities relative to key forage species.

- Develop multi-species management models to manage prey-predator systems effectively and optimise yields.

- Develop and implement conservative management approaches for new or expanding fisheries targeting forage species.

Localised overfishing conflicts

- Implement a co-operative state/federal effort to characterise the nature, extent and severity of localised overfishing conflicts.

- Refine data collection, fishery monitoring and law enforcement capabilities to support management actions within compressed time and area constraints.

- Develop and implement innovative management approaches that minimise or prevent localised overfishing conflicts.

Conflicts arising from fishing and gear impacts on aquatic habitat or protected species

- Institute a co-operative state/federal assessment to determine the types and severity of fishing and gear impacts on aquatic habitat and protected species.

- Develop a co-operative management and enforcement program to eliminate or minimise fishing and gear related habitat impacts.

- Encourage compliance through an information and education program.

Unethical fishing practices

- Work co-operatively with recreational and commercial fishing representatives to identify and accurately define illegal, detrimental "unethical" or bothersome fishing practices.

- In co-operation with the recreational and commercial fishing community, develop appropriate codes of ethics and educational programs needed to reduce conflicts.

- Stringently enforce regulations to eliminate illegal practices.

CONFLICT BETWEEN SHIPPING AND FISHING IN THE FINISTERRE AREA[1]

There has been exceptionally heavy shipping off Spain's Cape Finisterre (43°N, 9°W) for a long time, as it is the obvious route for all vessels leaving the main European ports and heading for the Mediterranean, South America and other points south. Traffic in the area increased as merchant shipping expanded, and the importance of North European ports and the growing scale of foreign trade made it one of the most congested shipping areas in the world. The increase in bulk oil shipments added to the problem.

Eventually a separation zone had to be established, dividing shipping into southbound and northbound traffic so as to prevent accidents in an area where weather conditions are often poor and fog and storms are frequent.

Fishing is also a major activity in the area, with boats setting sail from ports like La Coruna, Cayon, Malpica, Corme, Lage, Camarinas, Mugia, Finisterre, Muros, Portosin, Puerto del Son, Aguino, Ribeira and Marin.

Fishing grounds in this highly congested area are visited daily by trawlers or small fishermen. They include Boca de Corme, Mar del Crimen, A quiniela, Mar de la escoria, Puntal de Vilano, Pozo de Tourinan, Boca de Camarinas, Chan de Tourinan, Pozo de la nave and El profundo.

The fisheries reach to the edge of the continental shelf and down the upper slope located here between 14 and 18 sea miles off the coast stretching from Cape Vilano to Cape Finisterre.

Fishermen in the area use bottom trawls (mainly fleets from Muros, Ribeira and Marin), bottom longlines (those from Mugia and Anguino) and the *volanta*[2],

---

[1]   This paper has been presented by the Spanish authorities.
[2]   Bottom set gillnet.

which is becoming less popular. Many small *volantas* and *betas*[1] are still found inshore, however.

The leading commercial species in the area are blue whiting, horse mackerel, hake and Norway lobster.

As heavy merchant shipping traffic merges with vessels either fishing or making their way to and from the fishing grounds, a permanent risk of collision in fog or stormy weather hangs over the area, always the scene of unfortunate accidents and historically known as the "killer coast".

On 4th November 1993, following due preparatory work, the Assembly of the International Maritime Organisation passed a resolution adopting and confirming the location of a new Traffic Separation Scheme Off Finisterre, proposed by Spain, to enter into force on 4th May 1994 at 00.00 hours (see Figure 1).

The new scheme, which moves the separation zone further out to sea, is aimed at reconciling fishermen's rights and the rights of merchant shipping to use traditional routes, the ultimate objective being to improve safety at sea.

The ports nearest to the area covered by the new scheme are Finisterre, Mugia, Camarinas and Lage. Fleets using gear such as seines, *betas*, *volantas*, *rascos*[2], longlines, etc. are based in these ports and fish mainly in grounds on the shelf and slope to the north of Cape Finisterre (Figure 1).

The continental shelf waters at a depth of between 100m and 500m which were covered by the previous traffic scheme (Figure 1), also have fisheries that are regularly worked by trawler fleets from La Coruna, Ribeira and Muros.

---

[1]   Floating gillnet.
[2]   Wide-mesh bottom set gillnet.

Figure 1. **Traffic separation scheme off Finisterre**

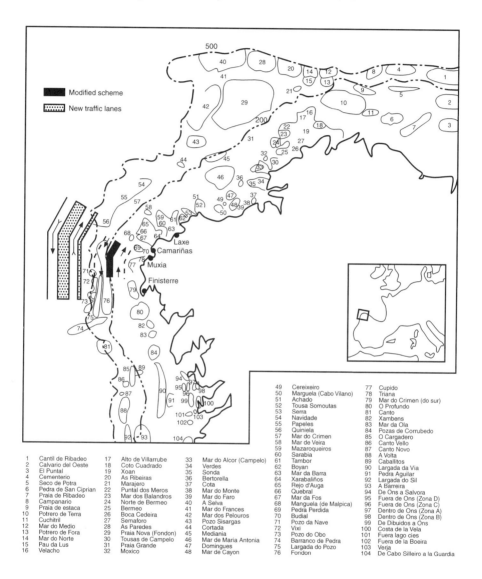

1	Cantil de Ribadeo	17	Alto de Villarrube	33	Mar do Alcor (Campelo)	
2	Calvario del Oeste	18	Coto Cuadrado	34	Verdes	
3	El Puntal	19	Xoan	35	Sonda	
4	Cementerio	20	As Ribeiras	36	Bertorella	
5	Seco de Potra	21	Marajero	37	Cota	
6	Pedra de San Ciprian	22	Puntal dos Meros	38	Mar do Monte	
7	Praia de Ribadeo	23	Mar dos Balandros	39	Mar do Faro	
8	Campanario	24	Norte de Bermeo	40	A Selva	
9	Praia de estaca	25	Bermeo	41	Mar do Frances	
10	Potrero de Terra	26	Boca Cedeira	42	Mar dos Pelouros	
11	Cuchitril	27	Semaforo	43	Pozo Sisargas	
12	Mar do Medio	28	As Paredes	44	Cortada	
13	Potrero de Fora	29	Praia Nova (Fondon)	45	Mediania	
14	Mar do Norte	30	Tousas de Campelo	46	Mar de Maria Antonia	
15	Pau da Lus	31	Praia Grande	47	Domingues	
16	Velacho	32	Moxico	48	Mar de Cayon	

49	Cereixeiro	77	Cupido
50	Marguela (Cabo Vilano)	78	Triana
51	Achado	79	Mar do Crimen (do sur)
52	Tousa Somoutas	80	O Profundo
53	Serra	81	Canto
54	Navidade	82	Xambens
55	Papeles	83	Mar da Ola
56	Quiniela	84	Pozas de Corrubedo
57	Mar do Crimen	85	O Cargadero
58	Mar de Veira	86	Canto Vello
59	Mazaroqueiros	87	Canto Novo
60	Sarabia	88	A Volta
61	Tambor	89	Caballitos
62	Boyan	90	Largada da Via
63	Mar da Barra	91	Pedra Aguilar
64	Xarabaliños	92	Largada do Sil
65	Rejo d'Auga	93	A Barreira
66	Quebral	94	De Ons a Salvora
67	Mar da Fos	95	Fuera de Ons (Zona D)
68	Manguela (de Malpica)	96	Fuera de Ons (Zona C)
69	Pedra Perdida	97	Dentro de Ons (Zona A)
70	Budial	98	Dentro de Ons (Zona B)
71	Pozo da Nave	99	De Dibuidos a Ons
72	Vixi	100	Costa de la Vela
73	Pozo do Obo	101	Fuera lago cies
74	Barranco de Pedra	102	Fuera de la Boeira
75	Largada do Pozo	103	Verja
76	Fondon	104	De Cabo Silleiro a la Guardia

SHORT HISTORY OF COASTAL ZONE PLANNING IN NORWAY, WITH EMPHASIS ON THE MUNICIPALITY OF VEGA[1]

Introduction

Due to its long coastline with a multitude of protected fjords, Norwegian coastal settlement and use of coastal zone areas have a very long and varied history. The use of the areas have progressed from a general extensive use, to the intensive use of specific localisations found today. An important cause behind this development is the past two decades rapid growth in the aquaculture industry.

Aquaculture differs from the traditional coastal zone activities like fishing and transport, through its exclusive demand for areas. A fish-farm excludes or largely limits other activities on and close to its location. As the aquacultural industry grew sharply in the 1970's and 80's, the need for a change in the legal framework regulating activities in the coastal zone became apparent.

A result of this process is that what is today known as "coastal zone planning", took form. The term coastal zone planning was introduced 10 years ago, and came as an answer to the needs for a more conceptual handling of the interactions between different interests in the coastal areas. Prior to 1985, there was limited legal framework designated to solving conflicts between different users of the coastal zone.

Due to the rapid expansion of the aquacultural industry, and the need for a legal base for solving conflicts of interests, the 1985 Planning Law (PBL) was approved. This law established the municipalities as the primary level for physical an economic planning. The municipalities were also given a legal framework to plan the use of the coastal zone out to the baseline- in

---

[1] This paper has been prepared by *Terje Doksroed,* Planning Officer, Regional Director of Fisheries, Bodø, Nordland, Norway.

correspondence with the existing right and obligation to plan for the land side areas.

The new law motivated municipalities to start work with the planning in the coastal zone. The motivation for this work can roughly divided into two categories; Settlement of conflicts, and planning for development;

- The need to solve conflicts of interest, was in large part limited to the southern and western coasts of Norway. As the aquacultural industry spread to the more sparsely populated areas along the coast, the need for coastal zone planning became apparent in these areas as well.

- Industrial development in coastal zone areas coincides largely with the localisation of aquaculture activities, and has also led to an increase in the need for planning even in the sparsely populated areas along the coast.

In the initial stages of coastal zone planning, two factors in particular caused problems in the planning process; There was a very strong concentration on aquaculture development, and a connection in the PBL between the right to plan in the coastal zone and the mandatory establishment of harbour districts.

- Aquaculture activities and economic development was the main reason for coastal zone planning in many municipalities, and the process in large part excluded other interests. Experience shows that this was a major reason why no municipalities succeeded in finalising their plans for the coastal zone areas prior to the 1989 revision of the PBL.

- The binding of the planning right to the mandatory establishment of harbour districts deterred many municipalities from starting work on coastal zone planning, as they feared economic consequences following the establishment of harbour districts.

During the 1980's, several new factors came to play an important role in public planning in coastal areas, and the further development of the planning process depended largely on these new aspects.

The increased competition for coastal zone areas led to a larger awareness for the work connected to protected areas and national parks. This was a new factor to consider for the planning authorities. Traditionally, protected areas has been very limited in the coastal zone. However, the increased pressure on the natural resources due to the development of tourism, leisure activities and aquaculture, and the general higher awareness for environmental problems, boosted this field of work.

The experience from the planning period also led to a new awareness of fisheries as a central topic in the formal planning of coastal zone areas. Traditionally, coastal fisheries with conventional methods has been a free activity in Norway. With regard to areas, fishing as an activity has been an integrated part of the development, and has for a large part provided the framework for other seaside activities.

As the development of the planning process has shown, this is no longer necessarily so, and it is therefore a need to implement coastal fisheries in the general planning process. This has created substantial methodical problems, partly caused by the cyclical character of the fisheries activity. Fisheries vary in time and character of area use, and most fishermen are dependent on the possibility of moving between fishing grounds in several municipalities and regions.

Fishing is regulated by separate laws, and is thus not directly involved in the planning process under the PBL. Fishing activity is nevertheless influenced by the designated use of sea-areas in the coastal zone, and it is therefore of great importance that fishing activities in the coastal zone is registered at an early stage, and that this information is used in the following planning process. Dispositions made under fisheries-law take preference over those made under planning law.

The legality of planning decisions made on municipality-level is monitored by regional authorities, this concerns particularly fishing activities and environmental issues. The regional level is the primary source of advice on formal and methodological issues for the municipalities. This co-operation is also prescribed in the PBL. In addition the PBL of 1985 also prescribes co-operation by the planning authorities with anyone, group, individual or administrative unit, that may be influenced by a plan.

The planning process in the Municipality of Vega

Vega is a coastal municipality consisting of the main island Vega, a number of smaller inhabited islands, and thousands of smaller islands and skerries. The sea around Vega is very productive, and for at least 10 000 years people have been living off the sea in this area. The municipality is situated in the southernmost part of Nordland county, in the Helgeland region, and the region is one of the major areas for aquacultural industry in Norway.

Archaeological excavations on the main island Vega has revealed settlements, fishing places and landing sites dated back to about 8 000 B.C. An interesting aspect is that the archaeologists claims that these revelations proves the existence of coastal zone "planning" at this very early stage. These findings represent important knowledge about the local history, and creates an interesting setting for the modern planning process in the area.

Vega was a very early participant in coastal zone planning in Norway, and started the planning as far back as 1984. The motivation for Vega and many other municipalities along the coast, was the wish to plan for an industrial development based on aquaculture. To map the possibilities for fish-farming, a survey on suitability of the coastal areas was conducted.

It early became evident that the limited concept first proposed for the plan would not be adequate. A clearly defined set of problems leading to a conclusive planning document found very little acceptance. On the contrary, it soon became evident that inclusion of as many parameters as possible in the preliminary phases of the process helped gain acceptance for the plan.

The 1985 PBL was revised in 1989, and the mandatory establishment of harbour districts were abandoned. This revision had little impact on the planning process in Vega, as the municipality established a harbour district at the very start of the planning process.

In Vega two main external factors have influenced the planning process.

• The New Plan for National Parks

• The LENKA project

The 1986 "New Plan for National Parks", included large areas in Vega, (the plan is at this time not completed in detail). In the areas covered by this Plan, aquaculture will probably be excluded, and other activities like fishing may also be influenced. The reason for proposing to   protect coastal zone areas are several; seabirds, otter, seals, botany, and landscape protection.

The planning in a municipality like Vega with limited economical potential, is obviously affected. On this basis, Vega entered into a co-operation with the National Nature Conservation Council with the objective to limit any negative consequences.

"The LENKA project" was an extensive survey to investigate the natural potential and limitations for aquaculture along the Norwegian coast. In Vega the LENKA methods and results were widely tested in the planning process. However, they proved not to be suitable for planning on municipality level.

The LENKA project has therefore had limited influence on the further work with coastal zone planning in Vega, and in other municipalities. One of the main reasons behind this is that the methods in the LENKA is based on topographical data in a zonal system as a basis for evaluating suitability for aquacultural purposes.

Experience in the Helgeland region of Nordland county has shown that the topographical characteristics indicating a low suitability in the LENKA context, often proves to be best suited location for fish-farms. The reason for this is the existence of archipels with moderate depths and strong current (tidal range 1.5 - 2.5 m), sheltering against the impacts of rough weather. The areas LENKA designated as best suited, have proved to be to exposed to rough weather, causing operational problems and high risk for damage to installations and fish. Nevertheless, the LENKA conclusion that Vega and the region has a very large capacity for aquaculture production is valid, with a negative match on the suitability criteria.

The fish-farming activity in Vega consist of two concessions for salmon farming, and a small number of concessions for other marine fish species and scallops. They are all localised in shallow protected areas. Environmental issues have been limited problems with seabirds and seals, which have been solved by technical adjustments to the installations (protective nets etc.).

One conflict that did arise, was the localisation of a fish-farm near to the run off from a land based fish-processing plant. This was resolved by moving the fish-farm, causing a limited drawback for the fish-farm as the distance to the landside base increased.

Some clarifications towards local maritime traffic was also needed. This was solved through minor adjustments of the placement of the fish farm, and thorough marking of the installations.

One experience drawn from the use of floating fish farms (-cages), is that each concession requires several sites. This is mainly necessary for separating different year-groups of fish, giving the best possible hygienic conditions. This means that the planning process must be able to adapt for the necessary development of existing concessions, as well as preparing for new ones.

There has been limited conflict between traditional fishing an aquaculture in Vega. In cases of conflict, the fisheries authorities tries to establish the degree of conflict, and to a certain degree the principle "first in time, best in right" is used to solve the conflict. Two kinds of fishing activities with a special legal status is important in this context; Net-casting sites, and Net-holding sites. For the last 25-30 years, these problems have had limited importance due to reduced herring stock, but as the herring fisheries is on the rise again, they have received renewed attention. Other sites important to fishing is also considered in the evaluation of aquaculture sites.

Through the planning process, Vega has uncovered a significant potential for developing aquaculture. However, many of the sites suitable for fish-farming are located within the proposed protected areas. The potential for economic development in Vega may therefore be limited.

In close proximity of the proposed protected nature areas, there is also a number of sites proposed for protection on historical and cultural grounds. Earlier several of the smaller islands were also populated, and traditionally these settlements "out-ports" close to the fishing grounds were important. The activity has been dropping steadily over the past 50 years and increased mobility of the fishing fleet has contributed to this trend. Recently more intensive fishing methods with longer stays at the fishing grounds, has created the need for temporary storage of equipment and fish. This has altered this trend, and created renaissance for some of the "out-ports". Fishermen and central authorities has thus found common interest in the preservation of buildings, piers, and other structures.

The growth of tourism, and the development of leisure activities based on nature resources have become an important priority for Vega. The possibilities for combining these activities with the protection of areas is also a central issue in the planning process, but as the work with protected areas is not finalised, the conditions for this combined use is not clear at present.

Conclusions

Coastal zone planning has been through a rapid development in methods and complexity. The attempts to transfer terrestrial planning to the coastal zone has proven that the challenges are different, and therefore requires new methods.

One aspect that had to be amended in the process, was the wish to obtain clear planning documents with an established framework and predicted use of different areas. Experience shows, that planning in seaside areas must be conducted in a very different context. Fishing, aquaculture and also to a large degree leisure activities, have proved to be of a much less static, and established nature than land-based activities. This emphasises the planning *process* itself as an important condition for the designation of areas to different purposes. This may be the most important experience of coastal zone planning so far.

The history of coastal zone planning is much shorter than the history of terrestrial planning. This means that the challenges remaining are many. These challenges are now processed in the ongoing work leading up to the next revision of the PBL. One clear conclusion is that a static plan is not suitable for the sea based activities in the coastal zone. The experience points towards the need to develop a more flexible legal framework to manage the planning challenges of the coastal zone.

This also confirms with the recognition that fisheries activities is best served with being regulated through an adaptive set of regulations as in today fisheries law. As there is a need to design specific areas to specific uses, one must always ensure that all relevant interests, including fisheries, are considered in the process leading up to the allocation of areas. However, the structured *process* of planning itself must be considered the most important aspect of coastal zone planning.

THE EFFECTS OF DUMPING WASTE FROM TITANIUM DIOXIDE PRODUCTION INTO THE HELGOLAND BIGHT[1]

When the authorities and the world of science discussed the possible consequences of dumping dilute acid into the Helgoland Bight, before the actual introduction of this policy in 1969, it was agreed that dumping would be harmless if certain precautionary measures were taken. In the administrative procedure which took place at that time it was however noted that supervision of the stocks of fish and micro-organisms on which they feed in the dumping area would provide proof as to whether harmful effects did actually appear in the ecosystem. As a result, an intensive monitoring programme was launched, the parties involved being the Federal Office for Maritime Shipping and Hydrography, the Bremerhaven Institute for Marine Research, today known as the Alfred Wegener Institute, the Biological Centre of Helgoland and the Federal Research Centre for Fisheries.

The Institute for Coastal and Freshwater Fisheries examined the condition of fish and micro-organisms on which they feed in the dumping area and surrounding areas. Since 1969, varying amounts of acid waste from titanium dioxide production have been dumped in the sea territory of the central Helgoland Bight 12 nautical miles north-west of the Island of Helgoland. The amount was between 750 000 tons (12 per cent sulphuric acid) and 450 000 tons (23 per cent sulphuric acid) per year. Characteristics of the waste were a low pH value and a high level of heavy metals. It was dumped from vessels of 1 000 ton capacity into the water in the vessel's wake at regular intervals during the trip, following established rules. The most important substance contained in the waste was, in terms of quantity, iron; since 1980 the amount dumped has decreased (40 000 tons in 1980, approximately 10 000 tons of iron in 1989). Among the heavy metals which are dumped in large amounts are chrome (approx. 70 tons were dumped in 1982), copper (213 tons in 1982), vanadium (190 tons in 1982) and cadmium (35 tons in 1982). In 1982 the so-called "green salt" was separated from the waste and no longer dumped. Since 1988 the amounts of heavy metals dumped have been reduced considerably.

---

[1] This paper has been prepared by the German authorities.

In the course of the investigations carried out by the German Hydrographical Institute increased concentrations of heavy metals were discovered in both the column of water and in the sediments near the dumping area. Chrome concentrations were found in the dumping area itself and in areas up to 150 nautical miles north of the centre of the dumping area.

The change in pH value during dumping in the area was negligible because of the high buffer capacity of the sea water.

In 1971 investigations into the possible effects of the dumping of acid waste on the numbers and distribution of fish showed no differences which could be linked to the dilute acid. In 1977 the investigations were repeated and these showed high percentages of externally recognisable diseases in the dab (*Limanda limanda*), a species of fish living at the sea-bed. This fish is landed in only very small quantities in Germany but plays a very important role in the whole ecosystem of the North Sea because of its massive numbers there. Three diseases were found to be particularly common:

- lymphocystis, a disease caused by viruses, also known in other species of fish;

- epidermal papillomas, skin tumours, cause unknown, possibly brought about by viruses;

- ulcerations, ulcers, caused by bacteria.

The frequency of disease found in the dumping area is subject to seasonal fluctuations. Therefore, lymphocystis and epidermal papillomas were found more frequently in summer in those years in which the investigations were carried out. During all study trips carried out in the dumping zone and the surrounding areas the incidence of epidermal papillomas was higher among dab than among fish from areas of comparison in the Helgoland Bight. These differences were very significant for the trips in spring. It was of no significance that fish in various areas of the Helgoland Bight were found to be of varying length. These findings were taken as an opportunity to test whether substances contained in waste, particularly chrome, appeared in increased concentrations in the sediments of the dumping zone and in the fish. It was discovered that the outer tissue of the dab in the dumping area contained higher concentrations of chrome than those in areas of comparison, and a link was found between the medium size of the epidermal papillomas on the fish and their chrome contamination level. These results, together with those concerning the contamination of the sediment, were passed on to the Federal Office of

Maritime Shipping and Hydrography, the licensing authority, indicating that more recent findings give cause for concern that the dumping is not without harmful effects.

Recently, in addition to externally visible diseases, the frequency of liver tumours among dab in the dumping zone and surrounding areas has also been investigated.

It was found that dab in the dumping zone suffered the most frequent occurrence of liver tumours, and this was not only within the Helgoland Bight but in the whole of the North Sea.

It was also found that the highest rates of anomaly in pelagic fish embryos of the species dab (*Limanda limanda*), plaice (*Pleuronectes platessa*), flounder (*Platichthys flesus*), cod (*Gadus morhua*), whiting (*Merlangius merlangus*), etc. occurred in the centre of the Helgoland Bight, i.e. congruent with the dumping zone for waste from titanium dioxide production.

These findings underlined the fact that demands made by the Federal Research Centre for Fisheries for an immediate cessation of dumping were justified. The publication of the results of the investigations led to critical debate at an international level. It was quoted in particular that, in all cases hitherto investigated, dumping had not caused any harmful effects. The procedure of Germany in giving notice of the end of dumping was regarded with scepticism. However, the results led to other countries also launching thorough monitoring programmes, the outcomes of which are now available. Notably, Dutch investigators used similar methods to test the condition of the fish and their contamination levels in the area where titanium dioxide waste is dumped off the Dutch coast. The most recent results show that in the Dutch dumping zone the occurrence of disease in dab is twice as high as in areas of comparison. This applies to externally recognisable diseases as well as to diseases of the liver. In addition, the outcome of the Dutch investigations indicate that dab in the dumping zone have higher levels of heavy metal contamination than fish in areas of comparison. The existence of these more recent Dutch findings emphasises the relevance of the work carried out by the Federal Research Centre for Fisheries. The results of the investigations led to an end to dumping of titanium dioxide waste in the North Sea as of the end of 1989. Direct disposal of titanium dioxide waste, particularly at the British coast, was continued until 1992.

The discovery of increased rates of disease in dab in the North Sea led to a change of opinion on the question of the sea's absorption capacity for harmful substances. Intensive supplementary investigations carried out by a variety of institutes have in the meantime revealed numerous ecological anomalies and thereby increased recognition of the fact that a reduction in dumping of harmful substances is urgently required.

After termination of dumping in January 1989, investigations into the incidence of certain diseases in the dab (*Limanda limanda*) and the appearance of deformed dab embryos in the former dumping zone and surrounding areas were continued. As in previous years, two trips were made, one in January and one in June/July, in order to record externally recognisable diseases and liver anomalies. Some results are given below.

Diagram 1 contains information about the fluctuations in terms of time in the incidence of lymphocystis in dab, a disease caused by viruses. Figures are recorded separately for summer and winter each year. If we observe the results recorded in summer, it is clear that for the years 1989-93 a constant decrease in the incidence of disease was recorded. An identical trend could not be found for the figures recorded in January.

Diagram 2 contains information about the variability over time of the incidence of liver tumours in dab (*Limanda limanda*), divided into groups according to size, > 25 cm and between 20 and 24 cm. All liver tumours greater than 2 mm were taken into account. In the years after termination of dumping, a constant decline in the relative incidence of this phenomenon in the livers of dab was recorded for both groups.

Diagram 3 contains information about the incidence of deformed embryos at all stages of development of dab in a coastal trip off the coast of Germany, Holland and Denmark in the period between 1984 and 1993. The data recorded in 1990, 1991 and 1992 come from Cameron and Berg (Morphological and chromosomal aberrations during embryonic development in dab (*Limanda limanda*). Mar. Ecol. Prog. Ser. 91: 163-169, 1992; Reproductiveness in fish. In: "Geht es der Nordsee besser?" Series SDN 1: 120-129, 1993). Maximum levels were found in 1987; more than 20 per cent of the dab embryos examined revealed anomalies. In the years after termination of dumping, lower levels of aberration were recorded in the whole of the area under investigation. The lowest value in the whole period of investigation was found in March 1993. On this trip it was also discovered that severe anomalies, which still occurred frequently in 1987 in the coastal strip mentioned, were now only found in isolated cases.

Caution is necessary when examining the more recent results of investigation represented in Diagram 3. It is known that the aforementioned diseases and anomalies cannot be caused solely by harmful substances. Other influences, such as hydrographical factors or quite generally severe changes in the living conditions of the fish, can lead to the same results. It should also be taken into account that, for various reasons, the situation caused by the contamination of this area of sea after 1989 has changed. In particular, deposits from the River Elbe have been reduced, and the amount of certain heavy metals has declined. The decline in the incidence of infection found here must therefore not necessarily be due to the termination of the dumping of dilute acid.

The investigations are being continued, and we will have to wait and see whether the trends becoming apparent here will be confirmed by future samples.

Diagram 1. **Incidence of lymphocystis in dab (Limanda limanda)**
Investigations in January and June/July respectively of each year in the former dumping zone for waste from titanium dioxide production

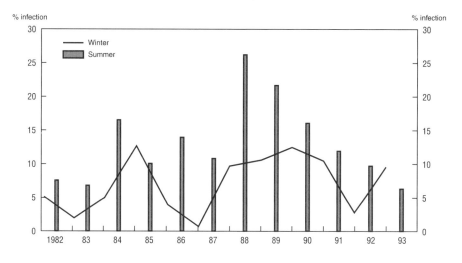

Diagram 2. **Incidence of liver tumours > 2 mm in dab (Limanda limanda)**

Investigations in January and June/July of each year in the former dumping zone for waste from titanium dioxide production

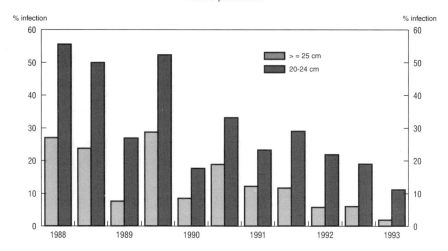

Diagram 3. **Incidence of anomalies in embryos of dab (Limanda limanda) at all stages of development found in German-Danish-Dutch coastal waters**

Values for 1990, 1991 and 1992 according to Cameron and Berg 1992; 1993

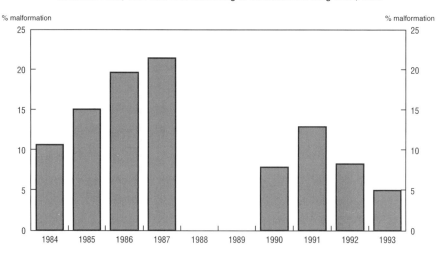

IMPACT OF DEVELOPMENT AND WASTE DISPOSAL IN THE COASTAL AREAS ON FISHING GROUNDS[1]

Damage caused on fisheries

Types of damage

Decline in productive capability of fishing grounds

Red tide: Red tide (akashio), generated by eutrophication as a result of excessive inflow of household and industrial waste water into the sea, is causing massive death for fish.

Oil pollution: Oil pollution occurs on coastal stocks and cultured lava as a result of oil spill caused by stranding of oil tankers, etc.

Impact of developmental projects on fishing ground environment: Loss of wetland and seaweed bed as a result of land reclamation in connection with various types of developmental projects in the coastal areas, including water-front development projects.

Damages to fish eggs and juveniles caused by massive water intake by power stations as well as impact on fisheries by rising water temperature in surrounding areas as result of outflow of warmed-up waste water.

Waste: Damage is caused on marine living resources by wastes such as discarded fishing nets, fishing line, and plastics and other substances discharged from households.

---

[1]   This paper has been prepared by the Fisheries Agency, Government of Japan.

Contamination of marine products and economic damage caused by excessive reaction of consumers

Contamination of marine products by toxic chemical substances: Contamination of marine living resources and retardation of their growth by toxic chemical substances such as mercury, PCB and organic tin compounds.

Shellfish poison: Poisoning phenomena and scallop and other species by poisonous phytoplankton, etc.

Oil pollution: Adhesion of oil and oil smell to marine products resulting from oil pollution by incidents.

Damage on fishing vessels and fishing gear, etc.: Damage to vessel cruise, fishing gear and fishing activities caused by discarded fishing nets, lines, and household wastes such as plastics and oil.

Scale of damage on fisheries

See following table.

Monitoring and prevention of pollution in coastal areas (General)

Monitoring and prevention of pollution on the basis of legal system

Pursuant to Article 15 of the Water Pollution Control Law, the prefectural governor monitors the contamination of water served for the public and underground water systems. In case of emergency, the governor can issue order for necessary measures to be taken by those who discharge waste water pursuant to Article of the same law.

Further, the Director-General of the Maritime Safety Agency conducts necessary monitoring of contamination of coastal areas pursuant to Article 45(1) of the Law Relating to the Prevention of Marine Pollution and Maritime Disaster. When the presence of significant marine pollution is identified, the Director-General shall, pursuant to Article 45(2), inform the head of the local governments, whose surrounding water is subject to contamination, of the situation.

Damage on fisheries caused by water contamination and other causes

	1987	1988	1989	1990	1991	1992
Total occurrences	313	294	285	295	251	233
Sub-total for marine fisheries	140	184	118	117	90	104
Red tide	47	40	22	42	30	31
Oil pollution by the hydrocarbonate	59	66	71	48	43	58
Other	34	28	25	27	17	15
Sub-total for inland water fisheries	173	160	167	178	161	129
Exhaust water from factories	38	27	24	31	35	28
Other	135	138	143	147	126	101
(Case for which damage value was identified)						
Total occurrences	107	79	94	73	71	61
Sub-total for marine fisheries	71	59	58	53	57	44
Red tide	24	17	6	15	22	9
Oil pollution by the hydrocarbonate	38	38	48	33	25	33
Other	9	4	4	5	10	2
Sub-total for inland water fisheries	36	29	36	20	14	17
Exhaust water from factories	9	2	6	5	4	1
Other	27	18	30	15	10	16
Total damage value (in million yen)	3 013	988	1 973	3 322	2 947	702
Sub-total for marine fisheries	2 938	931	1 782	3 288	2 918	657
Red tide	2 597	75	819	1 650	1 747	442
Oil pollution by the hydrocarbonate	116	350	737	1 249	580	196
Other	225	506	226	389	591	19
Sub-total for inland water fisheries	75	57	191	34	29	45
Exhaust water from factories	16	28	10	9	1	1
Other	59	34	181	25	28	44

Source: Data provided by the Fisheries Agency.

Monitoring implemented by the fisheries administration

Each prefectural government has its inspectors to preserve areas of importance for fishery production both in marine and inland water environment and also to monitor pollution. The inspector periodically monitors and investigates the pollution-caused damage in fishing grounds and extends guidance at those areas on ways to eliminate the damage and other emergency measures in case that damage occurs in fishing grounds. The Fisheries Agency subsidises for their monitoring activities implemented by them.

In areas where damage on fisheries is caused frequently by oil pollution and red tide, monitoring and preventive measures by means of vessels and aircraft are implemented by the Maritime Safety Agency and the Fisheries Agency, in parallel with activities by prefectural governments.

Specific countermeasures against pollution

Countermeasures against red tide

The main pillars of the countermeasures against red tide are:

- prevention of the occurrence of red tide,

- forecast of occurrence, and

- measures to prevent damage.

The Fisheries Agency charges research laboratories and other institutions with research and development of technology to cope with red tide. It also extends subsidy to prefectural governments for monitoring of the occurrence of red tide.

Prevention of the occurrence of red tide

Essential among countermeasures against the occurrence of red tide is reduction of total inflow of inorganic nutrients from land to closed sea areas. The Water Pollution Control Law was revised in October 1993 so as to regulate the discharge of total nitrogen and total phosphorus from industrial waste water for the purpose of the diminution of emission of inorganic nutrients from large-scale plants and other facilities as scheduled.

Also research has been underway on the technology to restrain growth of red tide-industry plankton and to reduce such plankton by means of micro-organisms.

Forecast of occurrence of red tide

As regards the area where the frequency of red tide occurrence is high, the Fisheries Agency is conducting extensive joint research on red tide in collaboration with fisheries laboratories of related prefectural governments, with a view to preventing the occurrence of damage on fisheries. Based on the findings of this research, the system of red tide occurrence is analysed to develop long term and extensive forecast model. Prefectural governments, for their part, are conducting research on the marine environment and monitoring survey of red tide plankton of specific places at the occurrence of red tide.

Furthermore, in 1991 fiscal year, the Fisheries Agency started building up a database of red tide-related information for prefectural sources.

Prevention of damage of red tide

With regard to prevention of damage of red tide on cultured fish and other resources, measures are taken to eliminate red tide plankton using hydrogen per oxide (H_2O_2) and acrinol, besides such means as suspension of feeding, transfer of aquaculture facilities and lowering the position of the cage.

A special aquaculture insurance system has been established as regards red tide damage as a relief measure for aquaculture operators in case of high mortality of cultured fish (e.g. yellowtail) and shellfish. When aquaculture is operated in the area designated under the Decree of the Ministry of Agriculture, Forestry and Fisheries, compensation is made from the insurance fund for the damage caused by red tide, on the condition that a special arrangement has been made in advance.

Countermeasures against shellfish poison

Shellfish poison is caused by shellfish eating poisonous phytoplankton. Sale and distribution of shellfish indicating values of diarrhoeic and paralytic shellfish poison exceeding the restricted level are prohibited under the Food Sanitation Law.

For this reason, the Fisheries Agency is taking the following measures, in collaboration with the Ministry of Health and Welfare, with a view to ensuring the safety of shellfish as food and maintaining stable supply of products:

- to establish for prefectural governments the procedures for monitoring of the occurrence of shellfish poison in the producing areas as well as the procedures for imposition (or removal) of voluntary restrictions on shipping in case shellfish poison value exceeded the restricted level; extend guidance to fishermen and other parties involved through notification and other means;

- to conduct nation-wide monitoring survey on shellfish poison;

- to implement research to clarify the shellfish poisoning mechanism, research on new types of shellfish toxin, develop techniques to measures the amount of shellfish poison and develop the methods of forecast of poison.

Oil pollution

As regards the damage by oil pollution for which the responsible party is identified, compensation for damage is made through negotiations among parties concerned because it pertains to civil proceedings. However, as regards the damage of oil pollution by unidentified cause, the public corporation placed under the supervision of the Central Government makes monetary compensation for the affected fishermen, and cover up the expenses equired for elimination of oil pollution from the fund collected from the industrial and other related organisations (the Central Government, prefectural governments and related sectors such as maritime, oil, machinery, electric power, fisheries).

Further, in order to minimise damage of oil pollution to fisheries, the Fisheries Agency extends to local governments the subsidy for purchase of oil fences, solvent for disposing oil, and other equipment to prevent oil contamination.

Impact of large-scale developmental projects on fishing ground environment

In assessing the impact of these projects on fishing grounds environment, the Fisheries Agency has launched development of the method to assess the importance of certain sea areas to fisheries, in particular, focusing on the preservation of coastal fishing grounds.

Impact of water intake and discharge by power stations on fisheries

In Japan, many of atomic and thermal power stations are built in the coastal areas, using sea water for cooling purposes. As a result, there is a large volume intake of sea water by those power stations, which causes in turn discharge of massive amount of warm exhaust water into the oceans. Because of concern about impact this massive intake and release of sea water may have on the fishing grounds, the Fisheries Agency is now conducting research at a number of power stations about the impact brought on the fishing ground environment by pumping-in of fish eggs and juveniles and by rising water temperature in surrounding areas.

Contamination by toxic chemical substances

Mercury, PCB and dieldrin

As mercury, PCB and dieldrin once presented a serious cause of pollution of fish and shellfish, provisional regulatory values have been established to ensure safety of marine products as food. Provisional regulatory values of mercury for marine products are set at 0.4 PPM for total mercury, and 0.3 PPM for methy mercury. The provisional regulatory values of PCGB remained in fish and shellfish are 0.5 PPM for production in distant water and 3 PPM for products in inshore and bay. Further, the provisional regulatory value for dieldrin remained in sea mussels is set at 0.1 PPM.

As regards the area where marine products harvested indicate values exceeding such regulatory value, the Fisheries Agency is extending guidance to fishermen to suspend voluntarily their fishing activities. The Fisheries Agency also monitors contamination status of fish and shellfish with special emphasis on these areas.

Organic tin compounds

As regards contamination of marine products by organic tin compound, no immediate danger to human health has been identified. But production and import volume of various types of organic tin compounds are regulated under the Law Concerning the Examination and Production of Manufacture, etc. of

Chemical Substances, with a view to preventing progress in contamination and improving the situation.

In addition, the Fisheries Agency is extending guidance to fishery organisations to refrain, as a principle, from using contamination-proof agents for fishing nets and ship-bottom paints containing organic tin compounds.

Disposal of waste originating from fishery activities

The Waste Disposal and Public Cleansing Law and the Law Relating to the Prevention of Marine Pollution and Maritime Disaster provide for the types of wastes to be disposed as well as their disposal methods.

Waste originating from fishing activities (e.g. discarded fishing vessels, fishing nets and shells), have been disposed of by individual parties who caused such waste. In recent years, in line with modernisation of fishing vessels and progress in aquaculture, occurrence of new types of waste is anticipated. These will include FRP fishing vessels and massive shells of oyster and scallop, disposal of which will exceed the capabilities of fishermen. In order to dispose these waste appropriately and preserve favourable fishing community environment, the Fisheries Agency has taken such measures as development of practicable disposal technologies for these waste, including recycling. The Agency has also been promoting establishments of disposal systems under fishermen's initiative.

Removal of sea-bed sediments in fishing grounds

Productivity of Japanese coastal fishing grounds is deteriorating as a result of accumulated sludge, resulting from inflow of industrial and household waste water and of soil and sand from land as well as floating plastics and other waste. Further, there have been inconveniences caused to fishing activities by these waste.

In order to restore productive capability of these fishing grounds, the Fisheries Agency is extending assistance to projects to remove sea-bed sediments such a sludge (by such means as dredging) and to retrieve plastics and other waste.

Education through public relations activities

The Fisheries Agency is promoting educational programmes targeted at fishermen and other parties involved to cope with pollution of fishing grounds by waste and is promoting campaigns to clean up fishing grounds and beaches.

PART II

COASTAL ZONE RESPONSES TO CHANGING FISHING POSSIBILITIES

SUMMARY AND OBSERVATIONS

Introduction

Coastal zone responses to changing fishing possibilities was the subject for the Committee for Fisheries' second part of its undertaking on coastal zone management. Three country case studies[1] were presented, and a number of Member countries contributed information on coastal zone activities in their country.

The coastal zone attracts an increasing number of people thus increasing the competition for space in the coastal zone and changes the relative importance between the coastal zone, cities and rural areas with regard to population and economic activity. In some countries, e.g. the United States, the coastal zone plays a particularly important role in the "territorial equation" because the number of people living there is important; the coastal zone accounts for 20 per cent of the land but contains 50 per cent of the population (and increasing).

The coastal zone is the home for a range of activities and is central to the fishing activity. The key issue in this Part II of the coastal zone activity is to highlight the linkages between the fishing industry and the other activities which take place in the coastal zone i.e. between the fishing industry economy and the coastal zone economy at large and to study policy responses to changing fisheries conditions in the coastal zone.

The coastal zone has an important role in providing alternative potential employment opportunities and sources of income for a fishing industry characterised by loss of fishing possibilities, low incomes and continuous

---

[1] Canada: "The Atlantic groundfish strategy"; Japan: "Revitalisation of fishing communities in Japan"; and Norway: "A description of the Norwegian fisheries policy objectives and the effects the measures have had".

restructuring, including contraction of fleets and a reduction in the number of fishermen.

It has often been stated that when the fishing sector is undergoing restructuring, an important element is to provide laid off fishermen with alternatives to fishing in the coastal zone. In fact the success of changes in the fishing sector will among other things depend on how easy it is to get fishermen into other occupations. The OECD's Committee for Fisheries, because of the characteristics of the fishing industry, has fruitfully contributed to a better understanding of linkages between the fishing industry and the coastal zone by identifying policy instruments which may help fishermen find alternative job opportunities. In fact several of the Committee's studies in recent years, on e.g. structural adjustment, management and economic assistance, have contributed elements, which have clarified these linkages and provided examples of policy instruments in use.

In many cases, a critical element in management plans which is often overlooked, is what to do with displaced labour. The low opportunity costs of most fishermen reflect the fact that they have few employment alternatives outside the fishing sector, hence their reluctance to leave the fishing sector.

The evidence

General comments

Changing fishing possibilities may come around for numerous reasons. The most cited example is stock depletion, but introduction of management measures such as licence limitation, ITQs, closed fishing areas and closed fishing seasons also impact directly on the fishing sector. While changing fishing possibilities may be a natural part of an industry which fluctuates, it does create pressure on the coastal zone economy, including demands for alternative job opportunities and may influence the overall economy of the area.

The country case studies submitted to the structural adjustment activity and the Review of Fisheries show that most Member countries are in a constant struggle to adapt the fishing industry to available resources. Common for most Member countries is that actions to adjust come late and consequently the adjustment burden becomes more difficult. It should be recalled that there may be a "natural" efficiency increase in fisheries due to technological innovations and that such developments may displace labour. In addition, late or slow adjustment of fleets to available resources may have serious socio-economic impact as well as resulting in low average incomes.

It is purposeful to distinguish between two main types of fishing communities in an OECD wide context i.e. the home bases for fisheries in developed areas where employment alternatives are available and where some infrastructure is present (harbour facilities, processing, and an active labour market) and remote fishing communities with no (or few) alternative employment opportunities and poor infrastructure.

In the OECD countries the lion's share of landings tend to concentrate around important port infrastructure. Modern fishing technology demands that services (ice, repairs) can be provided on the spot, i.e. are available in the harbours. Hence the activity of the segments of the fishing industry providing the largest shares of supplies occurs in areas where a certain infrastructure development is present and where the fishing sector is one of many activities.

As discussed in Part I of this report the developed coastal community is characterised by conflicts with other sectors of the economy (tourism, marinas, pleasure boating), and by one of encroachment. In such coastal zone areas the contribution of the fishing industry to the local economy is not big. However, it is from such areas where the largest supplies of fish are coming from.

Certain coastal areas in the OECD, although they may be small and isolated, are characterised by heavy dependence on fishing. The material made available to the Secretariat includes one comprehensive study on coastal zones dependency on fishing commissioned by the EU Commission. Covering the European Union, the Report ("Regional, Socio-Economic Studies in the Fisheries Sector", EU 1993) identifies fishing dependent communities in various ways according to the country concerned. Within the European Union 20 zones have been identified as heavy dependent on fisheries. Heavy fisheries dependent zones are located in Scotland, in the south and north-west Atlantic coast of Spain and east coast of Italy. Other examples, outside the European Union, include remote coastal areas of Norway and the Atlantic coast of Canada.

A problem encountered in remote fisheries dependent areas is a thinning of the population towards a "critical mass" level, where normal societal infrastructures like schools are at risk and/or where small/medium sized businesses cannot operate in a profitable way. Furthermore a "thinning" of the fisheries population makes the age structure skewed (e.g. Japan). Such developments further contribute to the problems of alternative employment opportunities.

Evidence provided by EU country case studies

Focusing on job creation initiatives as an adjustment help in the coastal zone context, the study undertaken by the European Union entitled "Regional, Socio-Economic Studies in the Fisheries Sector" (EU, 1993) provides the following evidence.

In the Netherlands there is no sector specific re-employment policy. There is a general re-employment scheme with a focus on re-educating unemployed to jobs in other sectors. The study concedes that fishing communities live isolated from other sectors of the economy, thus making the general policies inefficient vis-à-vis this particular sector.

In the United Kingdom (Scotland and Northern Ireland) fishermen, due to the payment structure with shares, are regarded as self employed. As such fishermen do not receive any unemployment benefits when laid off definitely; this is considered a clear disincentive to reduce capacity through decommissioning. The report suggests using the same type of schemes as were used when restructuring the steel and coal industry. This included the setting up of job shops helping unemployed finding prospective employers, writing application etc. A couple of re-training projects have been undertaken, where the success has been largely a function of the degree to which previous skills are relevant to new jobs. The aquaculture industry has not been seen as a viable alternative job solution in the short to medium term.

On the contrary, in Greece aquaculture is seen as a viable alternative to fisheries and with a potentially rapidly developing sector (sea bass and bream) where retired fishermen should be able to find new jobs. Difficulties have been encountered withdrawing fishermen from fishing, unless through retirement.

In Ireland a survey of displaced fishermen showed that only a limited number of them could find work in non-fishery related activity within the coastal zone they come from. The fishing sector has potential for growth but lacks capital and necessary qualifications for a modern fishing sector. There is no reconversion scheme specifically for fishermen.

The study on Denmark identifies a number of factors which are important in easing the reconversion of fishermen towards alternative jobs. Besides the presence of alternative industries within or outside the local community an important factor is the mobility (geographical and professional) of the affected fishermen. The study concludes that the coastal communities likely to be most affected by a reduction in fishing possibilities are the ones where these factors are not present or present at low levels. Hence such communities need help in the adjustment process.

On the EU level the European Regional Development Fund and the European Social Fund may co-finance (with Member States) infrastructure developments which may be needed for alternative industries, training costs etc.

Evidence provided by country case studies to this activity

The Canadian case study for this activity, as well as the Canadian case study submitted to the Committee's activity on restructuring the fishing industry, provide valuable information on how a concerted action in the wake of a closure of a fishery can be laid out and implemented. The Northern Cod Adjustment and Recovery Program, the Atlantic Groundfish Adjustment Program and the five-year Atlantic Groundfish Strategy complement each other with a view to providing substantial support for a development of alternatives to fishing. Training, job development, community development etc. are parts of this package adjusting the livelihood of up towards 30 000 fishermen and plant workers affected by the closures of the groundfish fisheries.

The programmes were effective in realising a better match between effort and stocks through withdrawal of licences (with financial incentives) and through early retirement from the fishing industry. The success of these programmes in addressing the more fundamental, long term, problems of adjustment, i.e. to achieve a better education/skills levels, provide job alternatives and a willingness to leave traditional ways of living in coastal communities, cannot be adequately assessed at this point in time.

The Canadian country case study notes that one reason behind the "stickiness" towards change of coastal fishing communities is that fisheries have evolved as an integral and traditional part of the social fabric of coastal communities rather than an self sustained industry. In turn this would suggest that any concerted policy action towards coastal zone/fishery problems should play on many levels including an appreciation of the functioning of the management system in place.

In a recent publication, "Regulation. Industry Structure and the Future of the North Atlantic Fishing Structure" (Doeringer et al. in No. 22, June 1995 of Canadian-American Public Policy), similar reflections are made i.e. "Institutional sources of labour stickiness are reinforced by the underdevelopment of many fishing communities...... If alternative employment prospects for the fishing industry labour were to be improved, there would be an increased incentive for surplus labour to leave the industry, the social opportunity costs of fishing labour would increase, the amount of sticky labour would be reduced, and optimal catch levels could be adjusted accordingly."

It is however conceded that for most Atlantic fishing communities little progress towards reducing labour stickiness has been achieved.

The issues raised in the previous paragraphs could help explain why fishermen and their organisations often are against changes to management systems which would result in a lower labour force (and particularly often when individual quota systems are proposed).

In Japan, due to the skewed age distribution of fishermen, revitalisation is important to sustain fishing communities for the years to come and induce younger people to enter the industry. To this end, the community based fisheries management system is being promoted where activities other than fishing activities are present and income from such activities can complement fishery income. In the Japanese contribution various activities are suggested, including:

- management of recreational facilities;

- management of marinas and yacht harbours;

- management of inns and private guest houses;

- direct sales of fish and fish products to tourists;

- seafood restaurants;

- guide service for recreational fishing;

- guiding and supporting recreational clam fishing;

- diving services;

- organisation of special events.

The Japanese case study entitled "Revitalisation of fishing communities in Japan" is particularly interesting as it provides suggestions for local employment initiatives. The paper furthermore provides a critical assessment of cases of success and cases of failure, including an identification of reasons for failure or success.

A country where the approach to coastal zone problems is particularly advanced is Norway where the Government is intent on contributing to equitable living conditions in all parts of the country. A stable residential pattern is considered important and requires regional and local growth centres. The country case study by Norway underlines the importance of the quality and availability of transport and communication services.

The Norwegian case study to this activity furthermore underlines the link between a sound fishing and aquaculture base and the possibilities for fostering and nurturing new activities in the coastal zone. In this regard the allocation of fishing quotas between various segments of the fleet plays an important role. The coastal fleet (as opposed to e.g. factory vessels) is labour intensive. When allocating quotas to the coastal fleet employment objectives for the coastal zone are addressed.

Evidence provided to the activity "Restructuring the Fishing Industry".

Using the information from the Committee's undertaking on "Restructuring the Fishing Industry" Member countries give varied impression as to the difficulty of alternative employment in the coastal zone.

In Australia it is recognised that short-term socially adverse consequences will be the result of a reduction in fishing capacity. However, if adjustment takes a long time, experienced fishermen are likely to retain employment. Certain regional communities that rely heavily on the fishing industry will be adversely affected.

In Finland employment in sea fishery has been decreasing for a number of years. This development is not entirely due to the stock situation, which with regard to herring and sprat is very good in the Baltic. The reason for the decline is mostly due to larger and more profitable units. Also reared salmon stock is abundant, but because of scarce wild salmon stock, this fishery is heavily regulated resulting in loss of employment of some fishermen. Aquaculture is regionally important providing employment opportunities and besides the main activity which in Finland is production of rainbow trout for consumption, it also contributes both to coastal and freshwater areas by enhancement of fish stocks and hence to fishing and tourism.

The contribution by Portugal considers it important to address:

- the retirement of older fishermen through pre-retirement schemes;

- reconversion of fishermen;

- improvement of qualifications though courses;

- creation of collective units for management and marketing;

- incentives to set up co-operatives/partnerships.

In the Netherlands fisheries in the macro-economic context provides few jobs, but in the typical fishing villages fisheries are a major contributor to the local economy.

Contrary to this situation both Iceland and Belgium report having difficulty recruiting to the fishing (harvesting) industry.

In Japan the fishing industry contribution to the overall economy cannot be ignored, creating employment for 1.7 million people. The Japanese study reports that from a regional point of view, and in particular where alternative employment is limited, the role of the fishing industry is crucial. The same applies to the home bases of the distant water fishing industry, which sustains dynamism to the local economy, as this part of the fleet requires large investments and advanced production systems.

Large decreases in the number of persons engaged in fisheries in Japanese local communities dependent on distant water fisheries were experienced through the 80's. It is reported that former long distance fishermen had difficulties in finding employment in the fishing industry; most changed occupation and moved to urban industries. The Japanese Government took special measures towards the long-distance fleet fishermen reductions which included circulation of vacancy notices, vocational retraining and unemployment insurance. It is, however, noted that former fishermen tend to be geographically and occupationally immobile and that new job alternatives outside the fishing industry may not be personally satisfying.

While the number of fishermen has decreased considerably in Norway the number of inhabitants in coastal communities dependent on fishing has only changed slightly. An active fishing industry, although not by sustaining artificial high capacity levels, is considered important to sustain coastal communities. Through contributing to stable and reliable supplies of fish, aquaculture may contribute to job and income stabilisation in the traditional fish processing industry.

Norwegian policy aims at maintaining a dispersed settlement in coastal areas. Stable and guaranteed incomes, through measures such as unemployment benefits and retirement pensions are secured by the Government. Furthermore, recruitment to the fishing industry is an important longer term objective which needs a well-established and functioning coastal community.

Other

The United States, in its Alaskan fisheries, has introduced a so-called Community Development Quota for halibut and sablefish. The programme, vested with a part of the quotas from each fishing area, allocates quotas to fishing villages for use by fishermen in there. Quotas are tradable subject to restrictions.

Conclusions

Remote coastal fishing communities in some OECD Member countries are particularly dependent on the fishing economy, although it is often difficult to measure the degree. The majority of landings take place outside areas significantly dependent on fisheries and the linkage between the boats' landing and individual communities is not monitored. Consequently, if landings alone are measured it is possible to underestimate the economic contribution fisheries make to the port of origin of the fishermen or overestimate the availability of alternative economic opportunities.

Disregarding the reasons for adjustment, many Member countries place particular attention to the fact that fishing communities are very slow to adjust to new realities. The combination of family supported fishing ventures, low educational levels, low levels of information, the freedom of fishing in most cases without paying for the common property, the chance of getting instantly rich through a big catch and remoteness, make fishing communities a special case. Traditional policy measures towards labour adjustment and reorientation have little if any effect in fishing communities. The Canadian experience may be particularly important to share with other OECD Member countries because of the evidence on the difficulties and time it takes to adjust fishing dependent communities to new realities.

The contribution by Japan to this undertaking is particularly timely as it provides various possibilities for alternative employment opportunities which fishermen could undertake to complement fishing income, or to create alternative income than from fishing. Likewise some suggestions are forwarded in the EU study "Regional, Socio-Economic Studies in the Fisheries Sector".

The Norwegian case study has, on the other hand, provided insight into a fairly successful policy which has integrated fisheries management with coastal zone objectives. Among OECD countries the Norwegian policy is the only where employment and settlement goals in the coastal zone have played a key role for years (since 1960 when the fisheries policy framework was laid down).

During the first round of discussion in the Committee for Fisheries on this part of the coastal zone activity it was evident that pursuing a more encompassing approach in managing fisheries is necessary; while fish stocks need management, fisheries deal with people as well, and solving fisheries problems need to address issues related to the fishermen, the fishing communities and the development prospects for the coastal zone. There are important human and socio-economic dimensions which have to be addressed if fisheries policies are to be successful.

Maintenance of community life has been mentioned by some Delegates as an important societal goal although market economies may not prescribe such goals as efficient. Government intervention through allocation of resources and the use of administrative means (e.g. allocation of fishing quotas) is therefore necessary. In the meantime, a more global, encompassing and integrated approach to coastal zone management is needed if such objectives are to be successfully implemented.

Against the background of continuous overfishing and need for restructuring, work towards improving the understanding of the social aspects of fishing communities is even more important and urgent. The Committee for Fisheries' activity on living marine resource management has identified effects of management policies and measures, while the economic assistance activity has identified, inter alia, the linkages between assistance policies and management as well as effects of certain assistance measures. Another building block for a sound coastal zone fisheries industry would be the identification of relevant and workable labour adjustment programmes which are efficient in the context of the fishing industry.

Insofar the experience of other OECD Member countries concur with the above conclusions, there is need for further developing reflections on job creation in the coastal fisheries community context. Traditional employment policy responses are not working efficiently in the context of the fishing industry and a more appropriate mix of measures which better target the particularities of the fishing industry will have to be advanced. To this end the OECD's Committee for Fisheries will consider social aspects of fishing dependent communities as a follow-up to its undertaking on the coastal zones.

THE ATLANTIC GROUNDFISH STRATEGY[1]

Introduction

Beginning with the collapse of the northern cod stock in 1992, the Atlantic fishing industry has experienced devastating declines in major groundfish stocks which were once its lifeblood. These declines compelled the federal government to respond swiftly to alleviate the hardship faced by fishery workers and, at the same time, forced fishery workers to take stock with regard to their future in the industry.

In response to the further declines in Atlantic Groundfish stocks, the Department of Human Resources Development in co-operation with the Department of Fisheries and Oceans announced, in 1994, the Atlantic Groundfish Strategy (TAGS). TAGS is a comprehensive five-year, C$1.9 billion program designed to provide income support, training and retirement initiatives for the approximately 30 000 fishermen and plant workers who were displaced by moratoria on major Atlantic groundfish stocks.

The program replaced the Northern Cod Adjustment and Recovery Program (NCARP), and the Atlantic Groundfish Adjustment Program (AGAP), which ended on May 15, 1994. The announcement of TAGS followed several months of extensive consultation with provinces, fishermen's associations, unions, industry representatives, fishing community representatives and other stakeholders.

This paper will provide a brief history of recent Atlantic groundfish declines and government responses to the collapse of these resources with a focus on the current Atlantic Groundfish Strategy. In addition, the paper will provide a brief commentary on the effectiveness of these programs in addressing both the needs of individual Atlantic fishery workers, and of the industry itself.

---

[1] This paper has been prepared by the Canadian authorities.

The Atlantic groundfish crisis

Northern cod

The collapse of the northern cod stock, brought into sharp focus the need for adjustment in the Canadian Atlantic fishery (Figure 1). In response to the dramatic decline in northern cod, the Minister of Fisheries and Oceans declared in July 1992, a two-year moratorium on the harvesting of the resource.

Figure 1. **1992 groundfish closures**

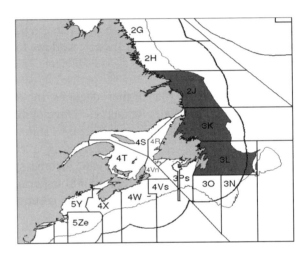

The Northern Cod Adjustment and Recovery Program (NCARP) was designed, first, as a bridging program to provide short term income support for over 25 000 fishermen and plant workers affected by the moratorium, and second to set the industry on a longer term course toward an economically viable fishery with a better balance between the number of individuals who depend on the resource and the ability of the resource to sustain them. The main elements of the program were:

- income support;

- training, counselling;

74

- early retirement;

- licence retirement;

- vessel support.

1993 Atlantic groundfish closures

It was not long after the introduction of NCARP that other Atlantic groundfish stocks started showing similar signs of decline. The 1993 Groundfish Management Plan reduced quotas in other major Atlantic Groundfish stocks by 60 per cent. At the time, scientific information indicated that the fisheries could continue and stocks would rebuild, provided that harvesting activities were conducted according to the conservation harvest plans announced for each fleet sector.

These quota reductions had the most severe effect where alternative fishing opportunities were limited, notably in Cape Breton, eastern Nova Scotia, northern New Brunswick, the Gaspé and the south coast of Newfoundland. An estimated 7 600 fishermen and plant workers faced lay-offs, or lacked sufficient work to qualify for Unemployment Insurance (UI) as a result of the quota reductions. Consequently, the Ministers of Fisheries and Oceans, and Human Resources Development in April, 1993, announced the Atlantic Groundfish Adjustment Program (AGAP) to address the needs of those who had either exhausted or who did not qualify for UI benefits. AGAP was comprised of elements that were consistent within those implemented under NCARP, i.e.:

- training;

- job development;

- transitional fisheries adjustment allowance;

- vessel support;

- community development program;

- fisheries alternative program;

- plant workers adjustment program.

In August of 1993, the condition of Atlantic groundfish stocks had further deteriorated. The Minister of Fisheries and Oceans announced the closure of five Atlantic groundfish fisheries, including those for cod stocks, on the eastern Scotian Shelf, in the southern Gulf of St. Lawrence and Sydney Bight, as well as cod and American plaice stocks off the south coast of Newfoundland (Figure 2).

Figure 2.   **1993 groundfish closures**

- 2J3KL Cod
- 3Ps Cod
- 4TVn (J-A) Cod
- 4Vn (M-D) Cod
- 4VSW Cod
- 3Ps American Plaice

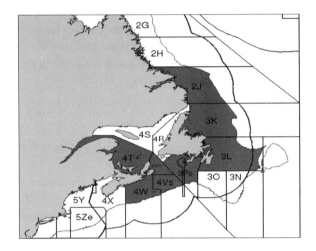

The closures, which were in addition to earlier quota reductions made in April, were announced in response to the recommendations of the Fisheries Resource Conservation Council and affected some 11 000 fishery workers. Extensive consultations with fishery representatives and the governments of the Atlantic provinces and Quebec were conducted to review the state of groundfish stocks and measures needed to improve AGAP to accommodate the increased numbers of affected fishermen and plant workers.

1994 Atlantic Groundfish Closures Under the 1994 Groundfish Management Plan, virtually all major directed cod fisheries had been suspended in all zones off Newfoundland, the eastern Scotian Shelf and the Gulf of St. Lawrence (Figure 3). Total Allowable Catches (TACs) for all species of groundfish in 1994 represented 25 per cent of 1988 levels. The Northern cod spawning stock was only five per cent of average. Other fisheries, such as pollock, redfish, silver hake, Greenland halibut and flatfish, continued on a limited basis with strict by-catch levels for cod and other restricted species.

Figure 3. **1994 groundfish closures**

- 2J3KL Cod
- 3Ps Cod
- 4TVn (J-A) Cod
- 4Vn (M-D) Cod
- 4VSW Cod
- 3Ps American Plaice
- 4RSPn Cod
- 3Ps Pollock
- 3LNO Haddock
- 3Ps Haddock
- 4TVW Haddock
- 4X Haddock
- 5Zjm Haddock
- 3M American Plaice
- 3LNO American Plaice
- 3LNOYellowtail
- 3NO Witch
- 3NO Cod

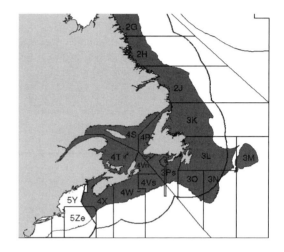

Effectively, the Atlantic groundfish industry was under a full moratorium. Scientific advice suggested it would be 6 years before a commercial groundfish fishery could resume in the northern Gulf of St. Lawrence. Similarly, would be four to five years for southern Nova Scotia. For northern cod, a minimum of 14 years would be required before the stock could sustain commercial harvesting of the resource (Figure 4).

Figure 4.   **Minimum recovery time for a commercial groundfish fishery**

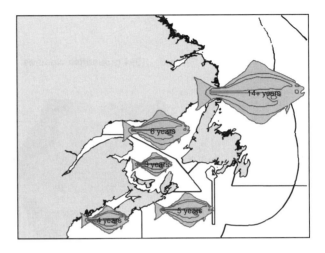

The Atlantic groundfish strategy (TAGS)

It was evident that the moratorium on northern cod in 1992 was merely a symptom of large scale declines in all Atlantic groundfish resources. The closures announced under the 1994 groundfish management plan resulted in the displacement of some 30 000 fishermen and plant workers. Unlike previous closures, however, these stock management decisions would have significant long-term impacts on Atlantic fishery workers and their communities where, in some regions, industry workers were confronted with the reality of a permanent loss of livelihood and way of life.

The Canadian Government was faced, therefore, with the challenge of mitigating the hardship caused by groundfish declines and, at the same time, continuing to promote adjustment in the industry. In this light, the Atlantic Groundfish Strategy was developed by the Department of Human Resources Development and the Department of Fisheries and Oceans. TAGS is a comprehensive 5-year program of adjustment and income support for displaced Atlantic fishermen and plant workers. The program replaces NCARP and AGAP which ended on May 15, 1994. TAGS has three major components: labour adjustment measures designed to help individuals in accessing new career training and job opportunities outside the fishery; measures to achieve capacity reduction and industry renewal; and the development of alternatives economic opportunities in the Atlantic region. The main elements of TAGS are described below.

Income Support

Income support payments are issued every two weeks and are based on individual average Unemployment Insurance (UI) benefit rates. The duration of support for eligible participants ranges from two to five years, based on the length of individual attachment to the groundfish industry.

Training

Income support and financial compensation are available for training activities such as: literacy training, community-based adult basic education, university study programs, leadership training, orientation/career guidance training and entrepreneurial training.

Green projects

Green projects provide a model and an opportunity for building linkages between employment-based programming and environmental action. The initiative responds to the stated needs of fishermen and plant workers to have the skills and job experience that will enable them to seize opportunities in the new economy, while ensuring responsible stewardship of the regions' resource base.

Employment bonus

An employment bonus is available to eligible participants who find and accept work outside the fishery.

Portable wage subsidy

A wage subsidy is provided to non-traditional fishery sector employers as an incentive to hire participants in permanent full-time employment and provide on-the-job training, as required. Assistance is also provided to participants to help them market themselves effectively to potential employers.

Self-employment assistance

Eligible participants interested in establishing a business are offered financial assistance to start their own businesses along with entrepreneurship training and technical support.

Community opportunities pool

Participants can become members of a community-based group (Community Opportunities Pool) to develop an inventory of projects and voluntary activities to be performed by members of the group. Project proposals and voluntary activities may come from the Pool itself or from any host or sector in the community: individuals, non-profit organisations, municipalities, other governments, and the private sector.

Participation requires a commitment of normally four to six months per year in an assortment of projects and activities of benefit to the community. Training is also provided.

Mobility assistance

Eligible participants are, on a voluntary basis, provided with financial assistance to relocate to areas where they can find work. The Employment Bonus and Portable Wage Subsidy incentives are also available. Participants remain eligible for support under the Employment Bonus or Portable Wage Subsidy components.

Youth

Special initiatives are available to help eligible youth prepare for new opportunities in order to make a successful transition from unemployment to employment, through literacy, work experience, training and mobility.

Fishery older worker adjustment program (FOWAP)

Long-service older groundfish plant workers aged 55-64 (or who will reach age 55 during their period of entitlement to benefits), are offered a voluntary retirement program in conjunction with and following negotiation with participating provinces.

Licence retirement program

The purpose of the Groundfish Licence Retirement Program is to reduce harvesting capacity through the permanent withdrawal of active groundfish licences. It is a voluntary program for those groundfish licence holders who meet certain eligibility requirements. Under this program, groundfish licence holders make their own assessment as to the worth of their licence, and submit a bid which is assessed against bids of other licence holders. If the licence retirement bid is accepted, the holder of the licence receives a payment in the amount specified in their bid. In return, they are required to:

- surrender their groundfish licence and their personal fishing registration;

- surrender other fishing licences they hold, or, where applicable, have them reissued in accordance with DFO licensing policy in their region;

- leave the commercial fishery permanently;

- forego any further TAGS income support benefits or TAGS adjustment options for which they may qualify.

In order to qualify for the Licence Retirement Program individuals must be TAGS eligible and also meet the following Special Eligibility Criteria (SEC):

- be the head of an active fishing enterprise;

- have seven full years of full-time fishing experience (five years with Full-time/Bona Fide registration status) and, either;

- had recent gross annual enterprise fishing income of at least $3 000 and 75 per cent of earned personal income from fishing in three of the four years used to qualify for TAGS; or

- had gross revenue from the fishing enterprise of at least $20 000 in two of the four TAGS qualifying years.

Early retirement

The objective of the early retirement program is to provide financial incentives for older groundfish fishermen to exit the fishery and to contribute to capacity reduction efforts. Those opting for early retirement will be required to exit permanently from the fishery by surrendering their personal fishing registrations and groundfish licences, and, where applicable, disposing of all other commercial fishing licences within a two-year period of receipt of early retirement. Non-groundfish licences not disposed of within two years will be cancelled permanently and not re-issued. Those who opt for early retirement will not be eligible to receive licence retirement.

The current early retirement proposal being discussed with provinces is a 70/30 federal-provincial cost-sharing. For consistency, the early retirement program proposed for fishermen will parallel the program delivered by HRD for plant workers. Income assistance payments would be calculated on the basis of 70 per cent of the individual's Unemployment Insurance benefits averaged over the three best years of 1988 to 1993, with a monthly maximum of $1 000 and a minimum of $630.

Effectiveness of fisheries adjustment and assistance programs

The primary goal of TAGS and past adjustment and assistance programs is to provide income assistance to fishermen and plant workers whose livelihoods have been virtually eliminated as a result of Atlantic groundfish declines. Additionally, and especially in light of current estimated stock recovery time lines, these programs were, and are, intended to promote adjustment in the fishery in order to bring about a better balance between harvesting and processing capacity and the sustainable limits of the resource.

NCARP was successful in achieving both objectives of the program. Original estimates were that some 25 000 fishermen and plant workers were affected by the moratorium and, of these, about 19 000 would require income assistance. In fact, 26 500 were eligible for the program and about 18 000 clients received income support. With respect to the early and licence retirement elements of the program, it was originally estimated that 700 fishermen would retire their groundfish licence under the licence retirement program and about 1 900 fishermen and plant workers would opt for early retirement. By the close of the NCARP program, 876 fishermen retired their groundfish licences and some 1 270 fishermen and plant workers exited the fishery through the early retirement incentive.

As successful as NCARP was, however, it did little to address the socio-economic conditions that persist across Atlantic Canada. A large portion of the training that was offered under the program consisted of literacy and numeracy upgrading. While this training is vital to fishery workers, it did not achieve the adjustment that was originally envisioned.   Adjustment is a long-term proposition in the Atlantic fishery. There are no quick-fix solutions to the problems of chronic excess capacity and over dependence on the fishery, stemming from the lack of economic alternatives; low education levels; and a steadfast determination on the part of many Atlantic Canadians to continue to pursue a traditional way of life.

The magnitude of the challenge became more apparent as the groundfish crisis deepened. TAGS consists of elements which were successful under NCARP (e.g., licence and early retirement), also it offers more adjustment measures, such as mobility and self-employment assistance.

However, as with the previous adjustment programs, the depressed education levels puts many of the active programming elements out of reach for a large portion of fishermen and plant workers on the TAGS caseload. For those who can avail themselves of the skills training offered under the program, the reality is that there are limited economic opportunities in the Atlantic region in which to practice these skills. Consequently, the only viable option for some individuals is to move to other areas of the country for employment. This can be a daunting and expensive proposition, not only for those who are faced with the decision, but also for the existing economies of the communities and regions in which they live.

Conclusion

The sudden and rapid decline in Atlantic groundfish resources, the mainstay of the Atlantic fishery, necessitated a swift response on the part of the Federal Government to address the needs of those individuals who rely on the fishery for their livelihood.   The programs were, and are, successful in responding to the immediate needs of affected fishery workers.   Income replacement was provided under each of the programs with strong incentives for individuals to acquire better skills so that they may be able to exploit other economic possibilities outside the fishery.   Licence and early retirement programs afforded many fishery workers the opportunity to leave the industry on a more solid footing and in a better position to seek alternative employment.

It is clear however, that these programs cannot be the only vehicle for adjustment in the fishery. The fishery has evolved as an integral part of the social fabric of many Atlantic communities rather than as a self-sustaining and viable industry. Consequently, the solution to the problems of the Atlantic fishery will involve not only the reform of fisheries management policies but also the development of strategies by all levels of government to address the socio-economic realities in the Atlantic regions - low education levels, depressed incomes and few economic alternatives.

REVITALISATION OF FISHING COMMUNITIES IN JAPAN[1]

Introduction

Fisheries in Japan is now faced with serious problems such as declining profitability, decreasing number of fishermen and an increasing proportion of aged fishermen.

Amid such a serious situation, in order to revitalise fishing communities diverse efforts in the fields other than fishery have been made by fishery co-operatives and their members, especially in the small coastal fishing communities. This paper is intended to specify patterns of revitalisation programs taken in those fishing communities and to evaluate the outcome from those specific programs.

Situation of coastal fishery and fishing communities in Japan

In Japan, "coastal fishery" is defined to comprise marine fisheries including fishing operation conducted with vessels of less than 10 GRT, set-net fishery, other fishing operations without using vessels as well as aquaculture.

Harvests by the coastal fishery segment in 1993 totalled 3 135 000 tons and Y 1 343.3 billion (37 per cent and 58 per cent of total marine fishery production). The number of people engaged in coastal fishery in 1993 was 275 000 which is 85 per cent of the total fishery work force. The number of vessels below 10 GRT operating in 1993 is 364 000, equivalent to 96 per cent of all the fishing vessels of marine fishery. These figures show that the coastal fishery holds an important position in Japanese fisheries.

---

[1] This paper has been prepared by *Yasuji Tamaki,* Management and Economic Division, National Research Institute of Fisheries Sciences, Fisheries Agency of Japan.

The average income of fishing household engaging in coastal fishery in 1993 stood at Y 6 592 000, a decline of 4 per cent from the previous year, within which the income from fishing activities accounts for Y 3 251 000, approximately 50 per cent of the total income.

The number of workers in coastal fishery declined 12 per cent over the past five years. Ageing is also a serious issue, with 39 per cent of people aged 60 years or older of the total male work force. It is a matter of urgency to secure younger work forces in order to ensure sound development of fisheries and maintain vital fishing communities in the long term.

There exist a total of 6 585 fishing communities in Japan and it can be summed up that there exists a fishing community every 5 kilometres along the Japanese coast by simple calculation with Japan's total coastline of 34 480 kilometres.

Further, the average number of fishermen per community was estimated at 42 based on the above-mentioned number, indicating that many small artisanal fishing communities are scattered along Japan's coastline.

Patterns of revitalisation effort in fishing communities

Revitalisation of fishing communities is essential for securing younger fishermen in the communities. Following are possible measures, other than fishing activities, to vitalise the fishing communities, which are taken by fishermen themselves. These measures do not include the ones aimed at directly revitalising fishing activities to promote "community based fisheries management" or those taken by the central or regional governments to revitalise fishing communities through improvement of fishing communities infrastructures.

Facility-oriented measures

This is the measure to provide facilities for the purpose to revitalise the fishing communities other than those used in normal fishing activities. These facilities usually require a large amount of investment.

Management of recreational facilities by fishing community

Sport and leisure facilities for promoting tourism are built utilising the natural environment where the fishing community is located. Besides facility fees, expanded consumption of local fishery products by tourists as well as expanded job opportunities for local residents can be expected. However, at the moment, there are only a few cases of this except for small facilities like piers for recreational fishermen.

As one of the typical instances of this type of leisure facilities, we can cite an aquarium a number of which have been established in the coastal area. However very few aquariums are run directly by fishermen, although fishermen usually provide fish to be displayed in those aquariums.

Management of marinas and yacht harbours

There are some fishery co-operatives which are running their own marinas for leisure boats.

Revenues from user fees for marinas are expected by fishery co-operatives through improvement of the port facilities and management of the marinas by fishery co-operatives themselves.

Although the number of privately owned yachts and motorboats was only 11 000 in 1965, it rose to about 161 000 in 1975, to 254 000 in 1985, and to about 320 000 in 1993. For this reason, a total of 203 fishing districts came to have marinas and other facilities in their district, in which there are 322 facilities in total where about 40 000 vessels are being moored. This category of activities has become a significant power which cannot be dismissed by the fishing community.

Management of inns and private guest houses

Some fishermen are running inns and private guest houses ("*minshuku*") in fisheries communities. Besides revenues from providing accommodations, they can add higher value to the fish they catch by providing them to their guests rather than selling them at the market. Further, job opportunities are created for their family members. There are cases in which guide services for recreational fishing or diving are offered at the same time.

Running inns or private guest houses has been one of the oldest part-time business of fishing households. Visitors look for family-type hospitality and fresh fish. Moreover, it is attractive for visitors that the accommodation charges are lower than those for ordinary hotels and inns.

However, there are some elements liable to keep tourists away from the use of these inns or private guest houses, such as communal use of bathrooms and toilets and difficulties to keep full privacy for rooms, etc.

The number of fishery entities engaged in the management of inns or private guest houses on a part-time basis in 1993 was 2 851, with direct employment of 3 573 people.

Management of direct sale outlets for fishery products

Fresh fish and processed fishery products from local catch are mainly sold at the direct sales outlets established in or near the fishing ports. This helps creating higher value-added to local fish products and expanding consumption, and also provide local residents with job opportunities.

There are many cases where fishery co-operatives are directly running such facilities.

In Japan, a trip to purchase low-priced fresh fish at such direct sale outlets at fishing communities has been a popular tourism activity. In some places, bus tours for such a purpose are organised. The number of fishing districts having such direct fish sales outlets in 1993 was 176, with 218 facilities in total attracting about 14 million visitors a year.

For communities wishing to attract many visitors to their direct sales outlets, it is important for them to be located in well-known fish landing sites with good transportation access from the town as well as large parking space.

Management of seafood restaurants

There are seafood restaurants designed to provide mainly locally procured catches. It is expected that these restaurants can enhance added-value of local fishery products, expand their consumption and also provide greater job opportunities to local residents. There are many cases where such restaurants are run by fishery co-operatives.

As seafood restaurants alone do not attract sufficient number of customers, they are often combined with tourism or recreational facilities as well as direct sales outlets.

Activities without specific facilities

Guide for recreational fishing

Guidance services for amateur fishermen to fishing ground are provided by using fishing vessels. There are two types in such services: one is providing opportunities for angling from the fishing vessels and the other is transporting clients by vessel to fishing points such as remote banks or small islands. In the former, they are classified into two types of services: (1) a vessel is chartered by a group of clients on an exclusive basis, and (2) individual customers share the same vessel. In both cases, fees paid by recreational fishermen is a revenue for the guides. Also those services are classified into two categories, a part-time basis and full-time basis guides.

Angling has a long history as a marine recreation, with a number of people making it their favourite pastime. The a number of young anglers has been growing, thanks in part to animation films on angling. As of 1993, the number of recreational fishermen in Japan totalled about 31.3 million (in the case of one individual fishing three times a year, the number was counted three), of which 11.1 million used boats and 20.2 million were shore-based fishermen. Of the anglers who fished from boats, 6.1 million received the service of recreational fishing guides, while 5.1 million did not use it. The number of those who, with their own vessels, do not use the service of recreational fishing guides is increasing in proportion to the increase of the ownership of pleasure boats.

The number of part-time fishing entities mainly engaging in recreational fishing guide business in 1993 were 3 212, and the people conducting recreational fishing guide were 20 987, of which 6 495 did it as their main business.

Recreational clam digging

This service is designed to allow tourists to engage in the catch of little clam and hard clam at seashore area in exchange for fees. Usually, fishery co-operatives release cultured juvenile shellfish in the areas under their management.

Clam digging is one of the oldest marine recreations in Japan, but natural beaches suitable for clam digging has decreased mainly due to reclaiming of coastal areas. On the other hand, construction of artificial beaches has come to increase lately. The number of people enjoying clam digging in 1993 was about 5 million in total (in the case of individual clam-digging three times a year, the number was counted as three).

Diving service

Some fishery co-operatives establish diving houses for recreational divers with shower and other facilities. Fishermen take tourist divers to the diving spots using their fishing boats. At some diving houses, several services are also available such as air tank rental.

Scuba diving has been increasing rapidly thanks to several factors such as the introduction of fashionable diving goods (e.g. introduction of colourful wet suits), improvement of goods for enhancement of safety (e.g. development of balancing compensator, emergency regulator) and improvements in instruments used (e.g. easier operations of water-proof camera). The total number of licenses issued by divers training organisations reached about 550 000 in 1993.

Organisation of events

Several events such as direct sales of the catch and fishery products aimed at expanding consumption of local fishery products and enhancing the publicity of the community itself are organised to attract more tourists from urban centres and neighbouring towns. Those events can also activate tourists' utilisation of guest houses in the region. The effect would continue not only during the event period but also after the event by enhanced publicity of the fishing community. Such events are usually organised jointly by fishery co-operative and municipal office or tourism office. Sometimes it is organised by the existing recreational facilities and/or fishery product direct sales outlet.

Besides the above-mentioned events, there are events organised by the young fellows association of fishery co-operative to promote friendships between youth in urban areas and fishing communities, sometimes with the aim to provide opportunities to young people in the community for finding future spouses, who are to succeed fishery business there.

Case studies

Successful cases

Fishery product sales outlet (UO-PORT, Kominato Fishery Co-operative, Chiba Prefecture)

Location conditions

Amatsu-Kominato Town is located within a distance of 100 kilometres from Tokyo (about 2 hours by train). The town has been popular as a tourist spot for many years. It has several tourist attractions including old Buddhist temples of Tanjoji temple and Kiyosumidera temple, and Tainoura (means sea of the sea bream) where tourists can enjoy first-hand sight of natural habitat of sea bream from pleasure boats. Moreover, it has three swimming beaches all of which are facing the Kuroshio (Japan current). Eating fresh fish procured locally is also one of the main purposes of many tourists coming to the town. The town has 22 inns and 83 guest houses (of which six are run by fishermen).

About 2.5 million tourists visited Amatsu-Kominato Town in 1993, of which more than half came by car. The number of visitors throughout the year is stable, with the average monthly number standing at around 200 000, with the exception of about 400 000 in August and about 100 000 in December.

The UO (which means fish in Japanese) Port is located near the entrance path to the Tanjoji temple as well as the port for Tainoura. Needless to say, this is considered most ideal since the main purpose of majority of tourists to Amatsu-Kominato is to visit cultural spots like Tanjoji temple as well as attractive spots like Tainoura.

Outline of the facility

The UO Port's direct sale outlet is one-story house with a space of about 400 square meters. A wide range of products are being displayed for sale, including fresh fish, fishery products and gift or souvenir goods. The facility was inaugurated in April 1992. Proceeds in 1992 totalled about Y 189 million and about Y 231 million in 1993.

Effects on regional fisheries

The UO Port is run directly by Kominato Fishery Co-operative. This project has been the most profitable one operated by them, with gross profit of about Y 44 million which is slightly more than a quarter of the overall profit of the fishery co-operative (Y 161 million). An ordinary profit of about Y 4 million was achieved only through direct sale project. As its overall ordinary profit was about Y 1 million, without the UO Port Kominato Fishery Co-operative might have suffered losses. Moreover, about 10 per cent of the flounder aquaculture production of the fishery co-operative was sold directly on a stable basis at the UO Port at higher prices than market prices. It can be concluded that the UO Port is considered to be one of the major support for their activities.

Fishermen have been able to prevent a drastic fall in fish prices as fishery co-operative can engage in direct sale of fish at the outlet where sizeable amounts have been sold. Furthermore, miscellaneous fishes which could not be sold because of its small quantity catch can now be marketed at the direct sales outlet. This has increased value-added of the fish. Moreover, employment opportunities for 8 persons were created as clerks of the sales outlet.

Reasons for success

- There seemed no disadvantages on the part of the fishery co-operative and fishermen in launching the project.

- The project could attract visitors easily since the town, as an old tourist spot, used to have a considerable number of tourists throughout the year.

- The cost required for establishing the site was fairly low. No cost was needed for acquisition of the site as the land was the property of the fishery co-operative. Y 50 million out of the overall project expenditure of Y 85 million for the construction of the facility was covered by the subsidies of the Prefectural Government.

Management of marina (Hakkei Marina, Kanazawa Branch of Yokohama Fishery Co-operative, Kanagawa Prefecture)

Location conditions

The Kanazawa Ward of Yokohama City is located in the suburbs of the Tokyo Metropolis, slightly less than 50 kilometres from central Tokyo, a distance of one-hour train travel. Because of its good location, conflicts occurred since around 1965 between local fishermen and owners of yachts and motorboats, who live in Tokyo and its suburbs anchoring their crafts in the fishing vessel mooring area in the Hirakata Bay. In an attempt to find a solution to this conflict, Yokohama Fishery Co-operative sent research teams to several marinas, including those situated along the west coast of the US and launched in 1975 the management of marina as one of the co-operative's services.

Outline of the facility

The sea-surface mooring method in which mooring buoys were set in the bay was introduced. Besides, the facility also has floating wharves, with about 40 rowing tender boats and an administration office. In 1991, a total of 209 yachts and motorboats were moored in the marina. Annual user fee for the facility was set at about Y 10 000 per one foot of the vessel's length, which is cheaper as compared with other marinas in the neighbourhood. Revenues from this services totalled about Y 32.22 million in 1991.

Effects on local fisheries

The gross profit from this services in 1989 amounted to Y 26 million, or 12 per cent of fishery co-operative's overall gross profit of Y 209 million. Revenues from the marina in 1993 stood at Y 32.22 million against expenditures of Y 4.01 million. A net profit of 28.21 million was obtained, which showed a high return rate of 88 per cent, resulting in an outstanding contribution to the management of the fishery co-operative.

The project was originally intended to dissolve the troubles between the fishermen and pleasure boats' owners, with no direct implications on fishermen's income. In this regard, it can be said that the fishermen are enjoying double merits from this project.

Several meetings have been held where fishermen can exchange their views with users, and consequently users' understanding of the actual state of fisheries is improving.

Reasons for success

- This project originated from fishermen's desire to reduce the number of pleasure boats invading in their mooring places for fishing boats. Also the establishment of a marina caused no disadvantage to fishermen (e.g. no loss of fishing grounds). For these reasons, it was quite easy to obtain consent and co-operation from many fishermen.

- There was no need to establish an embankment and other facilities as the mooring area itself situated at the naturally calm water area, and therefore equipment investment was less costly when compared with introducing land facilities. Because of the low investment it was possible to maintain the user fee for the marina at a lower level than other institutions in the neighbourhood.

- It could attract a large number of users because of its proximity to the Tokyo metropolitan area with lower user fee.

Diving support service (Tomido Branch of Ito Fishery Co-operative in Shizuoka Prefecture)

Location conditions

It is located on the eastern coast of Izu Peninsula, about 140 kilometres or about 2 hours trip by train from Tokyo. As it is rich in hot springs and blessed with warm climate, scenic seas and fresh fish products, it has attracted a throng of tourists especially in summer season. The Izu Peninsula has been popular for divers from Tokyo and its suburbs, and in Ito City, five out of six communities including Tomido are accommodating divers with services in some way or other.

The Tomido Branch, above all, has the oldest history in making its beaches available for divers. It has started their services since the establishment of the Izu Marine Park Diving Centre in December 1964, which is one of the oldest pioneering diving training institutions in Japan managed by a private corporation. This area also has 24 private guest houses.

Meanwhile, in Ito city, five out of six fishery co-operatives merged in January 1994 into one co-operative, Ito Fishery Co-operative, with the former fishery co-operatives functioning as its branches. In order for close analysis, data used in this analysis are based on the data before the merger.

Outline of the project

At the beginning, as stated in the foregoing, the fishery co-operative allowed the use of adjacent sea to divers using services provided by the private corporation. It however launched a full-scale diving service as its own project in January 1991 with co-operation of the Izu Marine Park Diving Centre and others.

Project revenues consist of rental fees for the use of air cylinder, fees for boat service (in case of boat-based diving), fees for the use of shower, lockers and hot spring as well as parking fees and others.

The administrative works are commissioned to the Izu Marine Park Diving Centre, and the fishery co-operative does not send any of its staff to the diving project itself. The air cylinders are purchased from the Izu Marine Park and another private company, and rented to divers with commission for the fishery co-operative. In boat-based diving, fishing boats are used for diving services. Thirty four boats, namely 94 per cent of all the vessels owned by the members of the co-operative are provided.

Effects on local fisheries

In 1991 the fishery co-operative reported a total gross profit of Y 121 million, of which 34 million or 28 per cent was derived from diving-related projects. This was the second largest independent business promoted by the fishery co-operative. (The largest revenue earnings from its entrepreneurial activity was Y 42 million.) Thus diving business constituted an important part in supporting the management of the fishery co-operative.

Thirty out of 32 motor-equipped boats belonging to the co-operative's members were used for boat-based diving, where each vessel earned commission of about Y 1 million in average, and it can be concluded that the project provides effective part-time business to fishermen.

Reasons for success

- This area is located fairly close to the metropolitan area and has a relatively long history of accepting divers. This attracted a substantial number of divers (about 100 000 a year).

- Local diving shops co-operated with the fisheries co-operative in launching and managing this project.

- Fishermen also has the merit of gaining income by extending guiding services to divers for boat-based diving.

Cases of failure

Diving service (Usami Branch of Ito Fishery Co-operative, Shizuoka Prefecture)

An instance in which the regional community became split as a result of the introduction of a facility for marine recreation.

Location conditions

Usami is located 15 kilometres north of the aforementioned community of Tomido. The community has 76 private guest houses, the largest number in Ito City. It was also designated as a branch of Ito Fishery Co-operative after the merger in 1994. As in the previous example, the data used here represent those compiled before the merger. Diving service business by a private company was launched in 1989, following a contract concluded with Usami Fishery Co-operative in 1988.

Outline of the project

Diving service is provided by the diving shop owned by the private company, and the fishery co-operative receives fishery co-operation fees from divers through the diving shop. Vessels owned by the members of the fishery co-operative were rented for boat-based diving services, and the owner of the boat receives fees part of which are paid as commissions to the fishery co-operative. Revenues of the fishery co-operative from this service in 1991 totalled about Y 3 million, and the number of divers in 1992 stood at around 7 000.

Contents of the failure

Initially, boats belonging to the hand-line fishing union were used by rotation for boat-based diving service. The hand-line fishing union consisted of the owners of large-size vessels for alfonsino fishing and those who owned small-size squid jigging vessels. Diving service shop gradually refrained from using squid jigging boats because they were too small and time-worn. For this reason, finally, the hand-line fishing union was split into two groups, one of which co-operated with the diving shop and the other of which do not. This confrontation ended up in the dissolution of the hand-line fishing union. The conflict between those two groups has now come to an irreparable point. The group which had decided not to cooperage for the diving shop tried to introduce another diving shop, which has further complicated the situation.

Reasons for of the failure

- The number of owners who wished to provide their fishing vessels for boat-based diving services was too large in comparison with the actual number of divers.

- There arose a polarisation of benefit to fishermen accruing from the diving service because of the difference of their fishing vessels as well as other factors.

REFERENCES

"Report on On-Site Survey concerning Formulation of Guidelines for Measures related to Marine Recreation (by type): Fiscal 1994," The National Federation of Fishery Co-operative Associations, March 1995.

"Report on On-Site Survey concerning Formulation of Guidelines for Measures related to Marine Recreation (by type)," The National Federation of Fishery Co-operative Associations, March 1991.

"Report on the Survey concerning Promoting Village Amenities (How making of a beautiful village can be related to revitalisation of fishing communities.) The National Research Institute of Fisheries Sciences, Management and Economic Department, October 1992.

"Report on On-Site Survey concerning Formulation of Guidelines for Measures related to Marine Recreation (by Type)," The National Federation of Fishery Co-operative Associations, March 1993.

"Manual for Measures to Be Taken vis-à-vis Marine Recreation; Adjustment of the Use of Sea Surface with Scuba Diving," The National Federation of Fishery Co-operative Associations, March 1993.

A DESCRIPTION OF THE NORWEGIAN FISHERIES POLICY OBJECTIVES AND THE EFFECTS THE MEASURES HAVE HAD[1]

Introduction

In this paper the Norwegian objectives and measures in place to keep local fishing communities viable are described. Furthermore the means that has been used to meet the Norwegian policy, objectives and their effects, are addressed.

Fisheries dependent areas in Norway

The localisation of the fishing industry in areas along the coast with few alternative possibilities for employment, combined with the fact that fisheries generate significant activities in the local communities and their related centres, highlights the fishing industry's vital role as an employer and patterns of settlement in sparsely populated coastal areas.

In Finnmark approximately 85 per cent of county's fishermen and fish processing workers live in communities where more than 10 per cent of population is employed in the fishing industry, whereas 38 per cent of the population of Finnmark lives in such communities. With the exception of a few city centres, most other economic activity is directly derived from the fisheries.

Within the other fisheries counties like Troms and Nordland there are relatively small differences between the fisheries-dependent communities at the county level. The fisheries communities of Troms and Nordland are more dependent upon this activity alone than those in Western Norway. In South Norway Møre og Romsdal and Sogn og Fjordane are the two most fisheries dependent counties. In North Norway, the fisheries communities form larger regional units, whereas fisheries communities in the south usually constitute a part of larger employment regions with a more differentiated basis for economic activities and employment.

---

[1] This paper has been prepared by the Norwegian authorities.

The general conclusion is that the northern part of Norway, specially northern part of Troms and Finnmark county is the most fisheries-dependent, both as regards the number employed in the fishing industry in relation to total employment and as regards the ratio of employment in the fish processing industry to total industrial employment.

During the 1980's problems in the fishing industry and other industries resulted in a reduction in the population in several municipalities in northern Norway. This situation changed 3-4 years ago and at the moment the situation is very positive. The following table illustrates the development in population from 1970 to 1994:

	1970-1980	1980-1990	1990-1994
FINNMARK	+ 2.6 %	- 4.8 %	+ 2.8 %
TROMS	+ 7.3 %	0.0 %	+ 2.6 %
NORDLAND	+ 1.5 %	- 2.1 %	+ 0.8 %
MØRE OG ROMSDAL	+ 5.5 %	+ 0.9 %	+ 0.8 %
SOGN OG FJORDANE	+ 4.9 %	+ 0.7 %	+ 0.9 %
WHOLE COUNTRY	+ 5.2 %	+ 3.8 %	+ 2.3 %

Regional policies

For the Government, full employment is the primary target of economic policy. An active industrial policy is required to increase activity in the private sector. Through close co-operation between the authorities, the business sector and the social partners, a good basis can be developed for meeting the challenges ahead.

Norway has a long tradition of supplementing private capital by a public engagement. The supply of public capital may help to finance projects with high risks or positive indirect effects. A Government engagement may also be important to ensure a flow of capital to vulnerable regions. It is the government's policy to maintain its engagement in private sector financing and to make it more efficient through the Norwegian Industrial and Regional Development Fund.

The Government is intent on maintaining the main features of the regional residential pattern, and contributing to equitable living conditions in all parts of the country. A reasonable distribution of employment and welfare opportunities

is a condition for achieving this aim. A stable residential pattern requires regional and local growth centres.

It is an aim to secure the best possible harmonisation between regional aims and aims related to other sectors. The Government feels that delegation and maintenance of target- and result-oriented controls are important in agriculture, the fisheries, the labour market and the educational sector.

The quality and the availability of transport and various communication services are of great importance to enterprises and households in the regions. The Government gives weight to this consideration in the formulation of transport and communication policies.

Tax policy will continue to be formulated with a view to giving special encouragement to industrial development in Northern Norway. Employers' social security contributions have been abolished in Northern Troms and Finnmark and reduced from 7,7 to 5,3 per cent in the rest of Northern Norway as from the beginning of 1993. In central areas in the south the employers' social security contribution is 14,3 per cent. Moreover, Northern Troms and Finnmark benefit from a special tax relief, and only the lowest rate of the so-called top tax is applied.

Fisheries policy

Norway, with its long coastline has through generations had access to large fish resources. The authorities are active in creating a substantial and economic fish industry and fish farming industry. Areas surrounding Norway are very productive and the fish industry therefore has good possibilities for creating both value added and employment along the coast.

The fisheries are one of the pillars of many local communities along the coast, and they still have an important development potential. The primary target of the Government's fishery policy is to create a sustainable and profitable fish and sea farming industry. Only a more profitable fish and sea farming industry can contribute to maintaining, and creating new activities in the coastal regions. The fisheries will continue to play an important role in the maintenance of settlement along the coast.

The Norwegian fisheries policy objectives can be summarised as follows:

- maintenance of the structure of settlement along the coast;

- conservation and protection of the living marine resources;

- maintenance of secure employment for the fishing industry;

- increasing the profitability of the fish industry.

The fish farming industry are mainly localised in coastal communities away from larger towns. Norwegian policy objectives for fish farming is to put considerable effort into research and development of new species for fish farming and examining the potential of sea ranching in Norway. The main objective is to provide new economic opportunities for those parts of the country where there are few alternatives fields of employment.

Resource management, regulations, etc.

Management of the fish resources is, in particular for north-Troms and Finnmark, of crucial importance for employment and settlement in the coastal areas. Changes in the fish stocks and quotas therefore result in large uncertainties in the employment within the fish industry, reduced profitability in the fish industry and this will also result changes in the structure of settlement.

Since 1960 more explicit regulation, both of effort and catches has been put into effect. Licence, quotas and other measures have been introduced as instruments to cope with the two most processing problems in the industry, deterioration of stocks and over-capacity. At present the Norwegian fisheries are regulated by a compacted and detailed set of laws and regulations. Most fisheries are today regulated with total quotas, quotas related to different groups and quotas related to each individual vessel.

The regulation of the fishing activities with regard to the allocation of quotas for different fishing gears and vessels reflects the need to take due account of biological considerations and the authorities' wish to achieve certain regional and employment goals. The coastal fleet is much more labour intensive than the seagoing fleet, and is therefore an important contributor to regional employment. Government regulations aim at a balanced composition of the fishing fleet. At the same time, the seagoing fleet is important from the point of view of covering the fish-processing industry's need for a supply of fish throughout the year.

Regional and employment considerations are also an important reason for the authorities' wish to limit the number of fishing vessels with on board production. These vessels are efficient and modern, and offer products of high quality. Nevertheless, this fleet will not in the longer run be able to ensure employment in the regions most dependent on fishing, although this has been the case to date.

A "fisherman-owned" fleet should be considered as a means to maintain also the capital of the industry in the regions. According to existing legislation, the permission to fish can as a rule only be given to active fishermen; and in companies the majority of shares or parts should be owned by persons fulfilling certain requirements with regard to activity. Fishermen-owned boats are considered as a guarantee of employment and settlement along the coast.

Legal basis

The fishing policy in Norway are based on several instruments. Some instruments are established by laws passed in the Storting, while others are taken at lower levels of the government. The background for decentralised decisions will be a law, usually formulated in a rather general way.

Three laws regulate the entry into the fishing industry:

- Act of 5 December 1917 relating to the registration and marking of fishing vessels. According to this act fishing vessels shall be entered into special register, and marked in a prescribed way.

- Act of 20 April 1951 concerning fishing with trawl. Under this law, a licence is needed for all trawl fishing.

- The Participation act of 16 June 1972. This law regulates the right to buy and sell fishing vessels. A general dispensation is given for vessels smaller than 50 feet, and special dispensation can also be given. This law also regulates the replacement of one vessel with another, the basic condition being that such replacement does not increase the catch capacity.

Experience has shown that input regulation is not enough to secure a sustainable use of the resources.

- The Act of 3 June 1983 relating to salt-water fisheries. This law gives the legal basis for a great variety of instruments used in management, in practice administered by the Ministry of fisheries, the Directorate of fisheries and other decentralised authorities in the day-to-day and year-to-year policy decisions.

Use of policy instruments in the cod fisheries in the northern part of Norway

In this paper we will focus on catch of Arcto-Norwegian cod because this activity is the most important, both in quantity and value. In 1994 about 68 per cent of the total first hand value in northern Norway was cod.

The catch capacity in cod fisheries can be put in to two main groups of vessels, the coastal fleet and the trawler fleet.

The coastal fleet is dispersed along the entire coast with a predominance in Northern Norway. Coastal fishing vessels, defined as vessels operating with conventional gear (nets, longline, hand line and Danish seine), are in general not subjected to licensing. Nevertheless, vessels longer than 90 feet need a special licence for operating with Danish seine. Main target species are cod, haddock, saithe and herring.

The cod trawler fleet is mainly located in the northern part of Norway. The trawler which catches fresh fish normally deliver their catches in local ports connected to fish processing plants in northern Norway. The most important target species for the cod trawler fleet are cod north of 62^0 N, commonly in combination with other kinds of white-fish, like haddock, saithe, Greenland halibut, redfish, etc. Since there is excess harvesting capacity in the cod fleet, no new licences have been granted the last 8-10 years. The catch capacity of the cod trawler fleet was in 1991 estimated to be slightly above 300 000 metric tons round/live weight pr. year.

In 1990, Norwegian authorities introduced a system where owners of cod trawlers were offered an extended quota until 1994 by merging two or more licensed vessels. A vessel was allowed to benefit from another vessel's quota, provide that this vessel is withdrawn from fishing on a permanent basis. This system are now considered in other parts of the fishing fleet.

Mostly, regulatory measures have been used in Norwegian fisheries. During most of the time these measures had a local and distributive character, e.g. reserving certain areas for specific gear. It is only in the last century that regulatory measures have been based on biological consideration. The first nation-wide regulations were mainly restricted to regulations on ownership of fishing vessels and of licences to fish. Only at a later stage, were regulation used explicitly to reduce fishing activity.

At times there has, through history, been tension between fishermen using old and new technology and different types of gear, (coastal fleet and the cod trawler fleet). As a consequence, trawling has been particularly subject to strong regulations. It has been argued that trawling is ecologically unsound by overexploiting young fish. Regulation of trawl fisheries has limited the area for this gear to a certain distance outside the cost, and landing of trawled fish in Norway has been restricted. The coastal fleet is fishing in the coastal zone area.

Today the cod fishery is regulated by national quotas. The quotas are distributed between different gears and between individual vessels. Until 1989, the coastal fleet was regulated by total catch quotas for cod for the coastal fleet as a group. This meant that the individual vessels fished for available resources within the maximum quota for that group. As a result of the severe reduction in catch quotas for cod north of 62^0N which took place in 1990, Norwegian authorities in 1990 introduced a new system for regulation of the coastal fleet, a system of individual quotas for each coastal fishing vessel. Quotas were granted to vessels with a minimum quota catch for cod north of 62^0N during the years 1987 to 1989. Among the vessels which were included in the system, the quotas were allocated relative to the length of the vessel. Vessels that did not qualify for a vessel quota had to fish within a specific group quota.

Because of its importance the coastal fleet has been allocated between 69 to 75 per cent of the Norwegian quota for cod in recent years. Within the limits of the group quotas of trawler fleet, quotas are allocated to each individual vessels. In 1995 the coastal fleet has 67 per cent of the Norwegian cod quota and the trawler fleet 33 per cent. The system is that the coastal fleet get a higher share of the total quota the lower the total quota is.

By dividing the fishing year in periods as part of the regulation of the cod fishery, it has also been possible to secure a dispersion of the activity along the coast so that different regions can use their comparative advantages and possibilities, depending on the availability of the cod. Since 1990 this system has been combined with vessel quotas in the coastal fishing fleet.

State aid

As a result of lower quotas for cod and haddock (i.e. a reduction of 60 to 70 per cent from 1987 to 1990) large parts of the fleet and processing industry experienced reduced income and liquidity problems. To maintain employment and to alleviate the liquidity problems NKr 100 million was made available as liquidity loans, of which NKr 75 million to the fishing fleet and NKr 25 million to the fish processing industry. Fishing vessels from regions dependent on fisheries where given priority. Factory vessels were excluded from the loans.

In 1990 special measures were introduced to alleviate the resource situation in the fishing industry and in the coastal regions. For vessels with loans or interest subsidies in the National Fishery Bank, subsidies were granted to cover parts of the interest on loans both in the National Fishery Bank and in commercial banks. In addition a reduction of interest on first mortgage loans in the National Fishery Bank was granted. Also the interest on building loans in the Norwegian State Housing Bank and the State Bank of Agriculture were subsidised in 1990 for fishermen from the five most northerly counties.

Since 1964, the Norwegian fisheries have received an annual support from the central Government, in accordance with the Principal Agreement for the Fishing Industry. On average the fishing industry received NKr 1.3 billion (in fixed prices) each year from 1964 to 1992. It is officially recognised that state support has been a major factor behind the over-capacity in the fleet.

Many of the most important fish stocks are in a process of growth and are expected to yield higher dividends in the next few years; exports of fish have increased steadily, and continued growth is expected in coming years. At the same time, government support to the fisheries has been reduced from around NKr 1.2 billion in 1990 to NKr 120 million in 1995. Thus in the course of a few years, the fisheries have developed into one of the leading growth industries, and constitute again a solid pillar of many coastal communities. The two remaining important Government support means are the minimum wage guarantee and the transport subsidies.

Also public financial support has been given for capacity reduction of the fishing fleet through scrapping and sales of vessels. A reduction in capacity will lead to higher catch quotas for each remaining vessel, which in turn will entail increased profitability for the industry as a hole and result in reduced pressure on marine resources. Fleet capacity has been reduced in the years 1990-1992, but further reductions will be necessary to increase profitability.

Investment support for building and modernisation of fishing vessels has also been reduced the last 8-10 years. The main reason for this has been the problem of over-capacity in the fleet and that too much investment support could increase the over-capacity. The Ministry of Fisheries lay down guidelines for which regions and what vessel types should be given priority. In 1995 more than 50 per cent of the investment support shall go to fishermen in Northern Norway. The investment support measures are administered by The National Fishery Bank. This bank also provide loan possibilities. A large part of the Norwegian fishing fleet is financed through this bank.

In addition vessels from Finnmark and Northern Troms have in the period 1989-1995 had an additional investment grant for building of fishing vessels and for purchasing vessels from other regions in Norway.

Fish farming industry

In Finnmark and northern Troms the need for special policy measures and development of employment opportunities is considerable. The fish farming industry is relatively new and limited in this area. Cold and dark winter seasons and an exposed coast are factors that limit the expansion of the industry in this region. High prices of smolt, high financial costs, considerably lower production per cubic meter than in the rest of country and losses due to disease, etc. lead to lower profitability in this region. The production of smolt is also low. Because of the need to create new industries in Finnmark and northern Troms, special policy measures to lead to growth of fish farming have been drawn up. But an improvement in the conditions of the fish farming industry in the northernmost regions also depends on the general conditions for this industry.

Effects of the means

It is difficult to exactly document the effects of the different means introduced. However it seems that the situation in the fisheries sector has an important impact on the situation in the coastal areas. The reduction in population in Northern Norway from 1980 to 1990 coincided with reduction in quotas and income in the fisheries in the same period. The same situation has been seen from 1990 to 1994 were we both see increases in population in Northern Norway and increase in quotas and income in the fisheries.

Employment in the fisheries sector has decreased every year since World War II. This is a natural consequence of rationalisation. In the county Møre og Romsdal 41 per cent of the fishermen are younger than 30 years old. In Finnmark only 24 per cent of the fishermen are younger than 30 years. In the north we have seen some difficulties with recruitment in the fisheries which result in ageing in parts of the fleet. This has probably to do with the structure of the fleet. It has been said that the strong offshore fishing fleet in Møre og Romsdal play a decisive role for the recruitment of young fishermen and that this factor is almost non-existing in Finnmark and North Norway. The structure of the fishing fleet in the north has been too one-sided. The reduction in number of medium sized coastal vessels, long liners, trawlers and purse seiners has made the regions more vulnerable for changes in the situations.

Until 1984 the coastal fleet could continue fishing for cod after the total quota was taken. Since then the fishery has been based on group quotas and maximum vessel quotas, and from 1990 also guaranteed vessel quotas for the most active fleet. From 1985 to 1989 the counties in Northern Norway faced a reduction in their share of the total cod catch. At the same time the two counties in Western Norway, Møre og Romsdal and Sogn og Fjordane, increased their catches. In 1989 and 1990 Northern Norway increased their share again, from 67 per cent in 1988 to 76 per cent in 1989, and except for Nordland county this share remained at a high level until 1994. Then the share was reduced to 69 per cent.

A report made by Nordland Research Institute conclude that the introduction of the individual vessel quota system in the coastal fleet in 1990 has not been only positive for Northern Norway. As an example Nordland's share has been reduced from 38 per cent in 1990 to 32 per cent in 1994. Finnmarks share has continued to decrease despite the fact that the regulation system the last two years has been adjusted in order to help cod dependent vessels and regions.

MAIN SALES OUTLETS OF OECD PUBLICATIONS
PRINCIPAUX POINTS DE VENTE DES PUBLICATIONS DE L'OCDE

AUSTRALIA – AUSTRALIE
D.A. Information Services
648 Whitehorse Road, P.O.B 163
Mitcham, Victoria 3132   Tel. (03) 9210.7777
                              Fax: (03) 9210.7788

AUSTRIA – AUTRICHE
Gerold & Co.
Graben 31
Wien 1   Tel. (0222) 533.50.14
                              Fax: (0222) 512.47.31.29

BELGIUM – BELGIQUE
Jean De Lannoy
Avenue du Roi, Koningslaan 202
B-1060 Bruxelles   Tel. (02) 538.51.69/538.08.41
                              Fax: (02) 538.08.41

CANADA
Renouf Publishing Company Ltd.
1294 Algoma Road
Ottawa, ON K1B 3W8   Tel. (613) 741.4333
                              Fax: (613) 741.5439

Stores:
61 Sparks Street
Ottawa, ON K1P 5R1   Tel. (613) 238.8985
12 Adelaide Street West
Toronto, ON M5H 1L6   Tel. (416) 363.3171
                              Fax: (416)363.59.63

Les Éditions La Liberté Inc.
3020 Chemin Sainte-Foy
Sainte-Foy, PQ G1X 3V6   Tel. (418) 658.3763
                              Fax: (418) 658.3763

Federal Publications Inc.
165 University Avenue, Suite 701
Toronto, ON M5H 3B8   Tel. (416) 860.1611
                              Fax: (416) 860.1608

Les Publications Fédérales
1185 Université
Montréal, QC H3B 3A7   Tel. (514) 954.1633
                              Fax: (514) 954.1635

CHINA – CHINE
China National Publications Import
Export Corporation (CNPIEC)
16 Gongti E. Road, Chaoyang District
P.O. Box 88 or 50
Beijing 100704 PR   Tel. (01) 506.6688
                              Fax: (01) 506.3101

CHINESE TAIPEI – TAIPEI CHINOIS
Good Faith Worldwide Int'l. Co. Ltd.
9th Floor, No. 118, Sec. 2
Chung Hsiao E. Road
Taipei   Tel. (02) 391.7396/391.7397
                              Fax: (02) 394.9176

DENMARK – DANEMARK
Munksgaard Book and Subscription Service
35, Nørre Søgade, P.O. Box 2148
DK-1016 København K   Tel. (33) 12.85.70
                              Fax: (33) 12.93.87

J. H. Schultz Information A/S,
Herstedvang 12,
DK – 2620 Albertslung   Tel. 43 63 23 00
                              Fax: 43 63 19 69

Internet: s-info@inet.uni-c.dk

EGYPT – ÉGYPTE
Middle East Observer
41 Sherif Street
Cairo   Tel. 392.6919
                              Fax: 360-6804

FINLAND – FINLANDE
Akateeminen Kirjakauppa
Keskuskatu 1, P.O. Box 128
00100 Helsinki

Subscription Services/Agence d'abonnements :
P.O. Box 23
00371 Helsinki   Tel. (358 0) 121 4416
                              Fax: (358 0) 121.4450

FRANCE
OECD/OCDE
Mail Orders/Commandes par correspondance :
2, rue André-Pascal
75775 Paris Cedex 16   Tel. (33-1) 45.24.82.00
                              Fax: (33-1) 49.10.42.76
                              Telex: 640048 OCDE
Internet: Compte.PUBSINQ@oecd.org

Orders via Minitel, France only/
Commandes par Minitel, France exclusivement :
36 15 OCDE

OECD Bookshop/Librairie de l'OCDE :
33, rue Octave-Feuillet
75016 Paris   Tél. (33-1) 45.24.81.81
                                   (33-1) 45.24.81.67

Dawson
B.P. 40
91121 Palaiseau Cedex   Tel. 69.10.47.00
                              Fax: 64.54.83.26

Documentation Française
29, quai Voltaire
75007 Paris   Tel. 40.15.70.00

Economica
49, rue Héricart
75015 Paris   Tel. 45.75.05.67
                              Fax: 40.58.15.70

Gibert Jeune (Droit-Économie)
6, place Saint-Michel
75006 Paris   Tel. 43.25.91.19

Librairie du Commerce International
10, avenue d'Iéna
75016 Paris   Tel. 40.73.34.60

Librairie Dunod
Université Paris-Dauphine
Place du Maréchal-de-Lattre-de-Tassigny
75016 Paris   Tel. 44.05.40.13

Librairie Lavoisier
11, rue Lavoisier
75008 Paris   Tel. 42.65.39.95

Librairie des Sciences Politiques
30, rue Saint-Guillaume
75007 Paris   Tel. 45.48.36.02

P.U.F.
49, boulevard Saint-Michel
75005 Paris   Tel. 43.25.83.40

Librairie de l'Université
12a, rue Nazareth
13100 Aix-en-Provence   Tel. (16) 42.26.18.08

Documentation Française
165, rue Garibaldi
69003 Lyon   Tel. (16) 78.63.32.23

Librairie Decitre
29, place Bellecour
69002 Lyon   Tel. (16) 72.40.54.54

Librairie Sauramps
Le Triangle
34967 Montpellier Cedex 2   Tel. (16) 67.58.85.15
                              Fax: (16) 67.58.27.36

A la Sorbonne Actual
23, rue de l'Hôtel-des-Postes

06000 Nice   Tel. (16) 93.13.77.75
                              Fax: (16) 93.80.75.69

GERMANY – ALLEMAGNE
OECD Bonn Centre
August-Bebel-Allee 6
D-53175 Bonn   Tel. (0228) 959.120
                              Fax: (0228) 959.12.17

GREECE – GRÈCE
Librairie Kauffmann
Stadiou 28
10564 Athens   Tel. (01) 32.55.321
                              Fax: (01) 32.30.320

HONG-KONG
Swindon Book Co. Ltd.
Astoria Bldg. 3F
34 Ashley Road, Tsimshatsui
Kowloon, Hong Kong   Tel. 2376.2062
                              Fax: 2376.0685

HUNGARY – HONGRIE
Euro Info Service
Margitsziget, Európa Ház
1138 Budapest   Tel. (1) 111.62.16
                              Fax: (1) 111.60.61

ICELAND – ISLANDE
Mál Mog Menning
Laugavegi 18, Pósthólf 392
121 Reykjavik   Tel. (1) 552.4240
                              Fax: (1) 562.3523

INDIA – INDE
Oxford Book and Stationery Co.
Scindia House
New Delhi 110001   Tel. (11) 331.5896/5308
                              Fax: (11) 332.5993

17 Park Street
Calcutta 700016   Tel. 240832

INDONESIA – INDONÉSIE
Pdii-Lipi
P.O. Box 4298
Jakarta 12042   Tel. (21) 573.34.67
                              Fax: (21) 573.34.67

IRELAND – IRLANDE
Government Supplies Agency
Publications Section
4/5 Harcourt Road
Dublin 2   Tel. 661.31.11
                              Fax: 475.27.60

ISRAEL – ISRAËL
Praedicta
5 Shatner Street
P.O. Box 34030
Jerusalem 91430   Tel. (2) 52.84.90/1/2
                              Fax: (2) 52.84.93

R.O.Y. International
P.O. Box 13056
Tel Aviv 61130   Tel. (3) 546 1423
                              Fax: (3) 546 1442

Palestinian Authority/Middle East:
INDEX Information Services
P.O.B. 19502
Jerusalem   Tel. (2) 27.12.19
                              Fax: (2) 27.16.34

ITALY – ITALIE
Libreria Commissionaria Sansoni
Via Duca di Calabria 1/1
50125 Firenze   Tel. (055) 64.54.15
                              Fax: (055) 64.12.57

Via Bartolini 29
20155 Milano   Tel. (02) 36.50.83

Editrice e Libreria Herder
Piazza Montecitorio 120
00186 Roma   Tel. 679.46.28
                              Fax: 678.47.51